Symbol

T0237971

M. Ferrari
I.-O. Stamatescu (Eds.)

Symbol and Physical Knowledge

On the Conceptual Structure
of Physics

Springer

Professor Massimo Ferrari

Università degli Studi dell'Aquila
Dipartimento di Storia e Metodologie Comparate
Via Roma 33
67100 L'Aquila, Italy
E-mail: paola.carrara@itim.mi.cnr.it

Professor Ion-Olimpiu Stamatescu

Forschungsstätte der Evangelischen Studiengemeinschaft (FESt)
Schmeilweg 5
69118 Heidelberg, Germany
and
Institut für Theoretische Physik
Universität Heidelberg
Philosophenweg 16
69120 Heidelberg, Germany
E-mail: stamates@thphys.uni-heidelberg.de

Library of Congress Cataloging-in-Publication Data.
Symbol and physical knowledge: on the conceptual structure of physics/M. Ferrari,
I.-O. Stamatescu (eds.). p. cm. Includes bibliographical references. ISBN 3-540-41467-3 (alk. paper)
1. Physics–Philosophy. 2. Signs and symbols. I. Ferrari, Massimo, 1954– . II. Stamatescu, I.-O.
(Ion-Olimpiu), 1941– . QC6.S917 2002   530'.01–dc21   2001020904

ISBN 978-3-642-07474-5

Springer-Verlag Berlin Heidelberg New York
a member of BertelsmannSpringer Science+Business Media GmbH

© Springer-Verlag Berlin Heidelberg 2010
Printed in Germany

Cover design: *design & production* GmbH, Heidelberg

Printed on acid-free paper

Preface

The question of the symbolic structure of physics is implicitly involved in any discussion about the character of physical knowledge and the development of physical theories. Actually many discussions would greatly profit from an *explicit* reference to and an investigation of this question, and much confusion may be avoided in this way. A book directly addressing the use and character of symbols in physics may help provide a point of view which is in the background of any consideration in the theory of knowledge and which is, to be sure, very relevant today, but – strangely enough – often seems to be missing as an explicit and central perspective in the present epistemological debates.

The concept of symbol has different meanings. Its wide diffusion in various cultural fields such as religion, mythology, art, and psychoanalysis, constitutes the best proof of its semantic variety, but also of the danger of using it in too vague a way. We start here from the concept of symbol conceived in the general sense of a sign or a material medium which is able, on the one hand, to communicate mental or conceptual contents and, on the other hand, to designate things or situations in the world. In this way symbols make it possible for human knowledge to "translate" the mental activity grasping the reality in conceptual frames, formal or natural languages, scientific propositions, theories and so on. There are obviously many different ways of interpreting the symbolical activity which forms the content of science; we shall try to exhibit a series of "case-studies", of epistemological reflections and delimited approaches aiming to show the complexity, but also the great importance, of the symbolization for that kind of human intellectual enterprise we usually define as "scientific knowledge".

This book is intended not to be a "loose collection of articles" but to represent a more coherent discussion concerning the symbolic character of physical knowledge. Since a full coverage of the subject was not a realistic goal, we have chosen a number of topics which we considered representative for the discussion, trying to achieve a complex but concise view of the problem. Thereby the main theme running through the book is the relation between the concepts used in science and the real world, a relation defining the symbolic character of knowledge in the natural sciences (for which the discussion about physics provides paradigmatic points of view).

The two chapters of the *Introduction* are meant to offer a brief overview of the subject in the frame of the modern history of the concept of symbol, and of modern physics, respectively.

The first chapter gives a short account of the history of the concept of symbol as well as of the main sources of the "cognitio symbolica", the philosophical-epistemological tradition starting in modern times with Leibniz. Logic, theory of scientific knowledge and language are the topics that may be considered as crucial for the development of this rich tradition, and philosophers such as Kant, Peirce, Schlick, Wittgenstein, Cassirer or scientists and logicians such as Helmholtz, Hertz, Boole, Schröder, Frege are discussed in order to offer the reader a concise, but useful historical overview.

The second chapter reviews the developments of modern physical theories, discussing features of their symbolic structure as they reveal themselves in the procedure of physics. Physical symbols are built in a stress field of conceptual (mathematical) relations and of references to "objects", identified in the encounter with the alien, external element at the empirical level. The author comments on the character of the physical symbols as it is shaped by these conditions and as it can be followed in the evolution of physical knowledge.

The three chapters of *Part II* concern the epistemological discussion on symbols and on knowledge about nature, as it occurs in the theory of science between philosophy and physics. They concentrate on two important points in this discussion, as represented by Duhem and Hertz, and on the relation to reality treated paradigmatically in the opposition Peirce–Cassirer.

Pierre Duhem is to be considered as one of the most important philosophers and historians of science who pointed out the significance of the role played by symbols in the construction of physical theories. His idea that physical knowledge – insofar as it forms a system – depends for a large part on symbolic constructions frequently led to the view that he advocated an instrumentalist or conventionalist conception of physical theory. In Karl-Norbert Ihmig's chapter it is argued that this view is mistaken or at least one-sided, since it overlooks the inner dialectic of Duhem's determination of the goal of physical theories being a natural classification of the phenomena. This dialectic consists in assuming an oscillation between the two poles of positivism and metaphysics. It is finally explained how this dialectic influenced Duhem's thesis of the continuity of scientific development with respect to a historical case study, namely the development of Newton's theory of universal gravitation.

In the second chapter of this part Andreas Hüttemann explains Hertz's concept of a symbol, or image, by relating it to this eminent scientist's central epistemological pursuit – the attempt to distinguish those features of our theories which are due to nature from those that we wittingly or unwittingly contribute ourselves. Hertz views his entire work in theoretical physics, both in electromagnetism and mechanics (not just the introductions to his books) as part of the attempt to solve this question. The concept of a symbol or image is a means he introduced in the *Principles of Mechanics* that allows him to draw ones attention to the ways in which various factors contribute to theory. The attempt of Hertz to develop a systematic approach of the symbolization question is the more relevant as it proceeds from "inside" physics.

In their decisive contributions to the philosophy of science of the 20th century Charles S. Peirce and Ernst Cassirer have developed different theories of symbol which have been highly influential in the philosophy of language, but still await a full evaluation of their relevance for natural sciences. While Peirce tried to transform the classical, post-chartesian theory of knowledge following Kant and Hegel into a semiotics of science, Cassirer used a critical recourse to Leibnizian philosophy to make the concept of symbol fruitful for modern physics. In his chapter, Enno Rudolph compares the conceptions of Peirce and Cassirer, against the background of the realism debate (d'Espagnat).

These three chapters therefore should provide insight into the general epistemological discussion concerning symbolization in science.

Part III is concerned with the discussion of the procedures of physics and of the character of physical theories and their development in the symbolic perspective. The five chapters of this part treat essential problems concerning the symbolic structures of physics: The well-definedness of the empirical decidability question in connection with discontinuous conceptual steps, which is an important moment in the evolution of symbolic systems (Martin Carrier); The role of intuition in physical thinking and its function in the symbolic constructions (Brigitte Falkenburg), and a systematic analysis of idealization as a basic procedure in these constructions (Andreas Hüttemann)– both of them providing essential elements for understanding the symbolization undertaking of physics; The discussion of the status of the quantum mechanical symbols, which makes explicit the acuteness of the questions related to the symbolic character of modern physical knowledge (Carsten Held); And the specific traits of the symbolic structures involved in the development of present day physics, with its abstract-mathematical and speculative-unifying character (Hans-Jürgen Pirner).

Martin Carrier focuses on the notorious problem of incommensurability. In his view, the symbolic character of scientific concepts becomes manifest in the dependence of concepts on the pertinent theory. The meaning of theoretical concepts is heavily intertwined with the substantial claims of the theory to which they belong. It follows that theoretical change may induce conceptual divergence – with the result that concepts corresponding to one another at first sight fail to be intertranslatable upon closer scrutiny. Incommensurability is portrayed as translation failure due to theoretical incompatibility. As a result, potential conceptual analogs either fail to preserve the conditions of application or fail to reproduce the relevant inferential relations. Carrier analyzes these features using the conceptual relations between classical electrodynamics and special relativity. He points out that incommensurability does not impair the possibility of empirical comparison and it is in fact directly related to the emergence of new knowledge.

Physical theories are said to be non-intuitive, for they are formulated in the abstract and symbolic language of mathematics. What counts as intuitive, however, is a product of historical processes. For an Aristotelian of the 17th century, the Copernican theory and the Newtonian mechanics were just as non-intuitive as the theories of Heisenberg or Hawking are for us today. In her chapter Brigitte Falkenburg sketches the traditional epistemological debates concerning language and reality of physics and shows how Kant's theory of intuition was supposed to close certain semantic gaps between the language of mathematical physics and our world of experience. She indicates, again with recourse to Kant, the extent to which contemporary physics, too, relies on intuitive concepts in order to embed its abstract system descriptions into natural language usage and how physical symbols are constructed in this procedure.

In his chapter Andreas Hüttemann attempts to classify various kinds of idealization, asking why physicists make use of them. He argues that physicists make use of idealizations in order, first, to discover those components of physical systems with the help of which the behavior of complex systems in nature can be explained and, second, in order to describe them such that they fall within the range of application of simple models. As a consequence Cartwright's conception of idealization and the conception that has been associated with Galileo have to be modified. He furthermore argues that the use of idealizations is direct evidence for Hertz's claim that physical knowledge is symbolic.

The analysis of Carsten Held is directed at the deep, structural difference between classical and quantum physics, contradicting the superficial impression that, say, states and events are symbolized in different, but in fact parallel, fashions. The probabilistic reading of the quantum-mechanical state vector, for instance, conflicts with its role as a state description. Quantum physics necessarily frustrates a certain ideal of physical description or modeling of the real world. The discussion shows that a classical ideal of intuitively accessible descriptions of states and events which governs our initial understanding of the symbols in a physical theory is at a stake in the epistemology of quantum mechanics.

Finally, the last chapter raises the question of whether it might not be useful to go beyond the concept of symbol and start from the general concept of sign in the context of a theory of communication and a theory of structure. This semiotic aspect is debated following the latest developments of physics: large scale computations, the theory of complexity and string theory. Hans J. Pirner asks whether it is useful to speak of a "postmodern" age of physics and whether we can learn more about the frontiers of physics by uncovering the semiotic and/or symbolic elements of a developing theory and phenomenology?

The five chapters of the third part therefore allow a closer view of the actual proceedure of physics from the point of view of symbolization, and of its significance for the construction of physical knowledge.

The discussion in the book continually reveals the necessity to correctly appreciate the symbolic character of knowledge in developing epistemological considerations. It attempts to provide an at least consistent, if far from comprehensive, view, which could also give incentives for more far-reaching programs.

The book emerged from a project addressing *Symbolization in Physics* conducted at *FESt (Forschungsstätte der Evangelischen Studiengemeinschaft – Institute for Interdisciplinary Research)*, Heidelberg, and although the various contributions retain their individuality they reflect the common work of the authors in the frame of this project.

The above enterprise was one of the interdisciplinary projects of FESt. It is the merit of this institution that it provides a solid frame for research in, and dialogue between, natural sciences, social sciences, humanities, ethics, and theology and can therefore promote long–term interdisciplinary research projects which would scarcely be possible elsewhere.

We highly appreciate the engagement of FESt in this irreplaceable activity and are greatful for its support of, and confidence in our project.

We are grateful to Wolf Beiglböck for competent advice and assistance in the completion of the book and to the Springer team for excellent copy-editing work and the preparation of the manuscript.

The special thank of the editors goes to the authors of this book in acknowledgment of their highly competent and motivated activity in the project.

Heidelberg   Massimo Ferrari
May 2001   Ion-Olimpiu Stamatescu

Contents

Part I Introduction

**1. Sources for the History of the Concept of Symbol
from Leibniz to Cassirer**
Massimo Ferrari 3

1 Leibniz's "Cognitio Symbolica" and the Theory of Expression . . . 4
2 Kant: Schema and Symbol 8
3 The Rediscovering of the Sign 12
4 Hermann von Helmholtz and Heinrich Hertz 16
5 Intermezzo: Hertz and Wittgenstein 20
6 From Moritz Schlick to Ernst Cassirer 22
7 Knowledge as Symbolic Form 25

**2. On the Use and Character of Symbols
in Modern Physical Theories**
Ion-Olimpiu Stamatescu 33

1 Some Remarks About Physics 33
   1.1 On the Physicist's Part in Epistemological Discussions 33
   1.2 On the Requirements upon Physics 34
   1.3 On the Dynamics of Physics Development 36
   1.4 Questions of the Dynamics of the Symbolic Structures 38
2 Some Remarks Concerning Physical Symbols 39
   2.1 Symbols and Things 39
   2.2 The Valences of the Physical Symbols 41
   2.3 On the Status of the Symbols 41
   2.4 On the Association of Symbols to "Things" 43
   2.5 The Constructive Aspect of the Necessity
       in the Association Symbols-to-Things 45
   2.6 On the Establishment of Physical Concepts 45
3 Modern Fundamental Physical Theories
   and Their Relations 47
   3.1 The Structure of Physics Research 47
   3.2 Established Physical Theories 50
   3.3 Relations Between Theories 52

3.4 Evolution of Symbols 58
3.5 On the Symbols of Modern Physical Theories 61
4 On Physical Knowledge and Scientific Progress 63
4.1 The "Pragmatic" Attitude of Physicists 63
4.2 On Justification and Truth 65
4.3 On Intuition and A Priori 67
4.4 On Scientific Progress 68

Part II   Views on Symbol in the Philosophy of Science

3. The Symbol in the Theory of Science: Duhem's Alleged Instrumentalism or Conventionalism and the Continuity of Scientific Development

Karl-Norbert Ihmig 75

1 Changes in the Theory Concept During the 19th Century 75
2 The Symbolic Nature of Scientific Theories 77
3 Duhem: An Instrumentalist or a Conventionalist? 80
4 Natural Classification and the Systems Concept 85
5 Natural Classification and the Continuity
of Scientific Development 91

4. Beyond Realism. Symbolism in the Philosophy of Science by Charles S. Peirce and Ernst Cassirer

Enno Rudolph 97

1 Sign and Science: Peirce and His Predecessors 98
2 Symbolism or Realism: Is There an Alternative? 104

5. Heinrich Hertz and the Concept of a Symbol

Andreas Hüttemann 109

1 Hertz on Helmholtz's Theory of Signs 109
2 The Comparison of Electrodynamic Theories 110
3 The Objective of *Principles of Mechanics* 113
4 The Concept of a Symbol or Image
in *Principles of Mechanics* 114
5 Images and Models 116
6 Criteria for the Evaluation of Images 117
7 Problems 119
8 Conclusion 120

Part III   On the Symbolic Structure of Physics

**6. Shifting Symbolic Structures and Changing Theories:
On the Non-Translatability and Empirical Comparability
of Incommensurable Theories**
Martin Carrier 125

1   Symbolic Descriptions and the Choice
     of Conceptual Structures 125
2   Meaning, Theoretical Context, and Adequate Translation 126
3   Shifting Theoretical Ground:
     the Example of the Einsteinian Revolution 130
4   Incommensurable Quantities in Classical Electrodynamics
     and Special Relativity 133
5   Incommensurability, Split-up of Natural Kinds
     and Shifts in Reference 138
6   Empirical Comparison of Theories
     with Incommensurable Concepts 140
7   Conclusion 146

7. Symbol and Intuition in Modern Physics
Brigitte Falkenburg 149

1   Language and Reality 150
2   Functions of Intuition 158
3   The Graspability of Cognition 166

8. Idealizations in Physics
Andreas Hüttemann 177

1   An Example of an Idealization 177
2   The Concept of an Idealization 177
3   Different Kinds of Idealization 178
     3.1   Production of Physical Systems 178
     3.2   Isolation 179
     3.3   Data Interpolation 179
     3.4   Data Fitting 180
     3.5   Abstraction 181
     3.6   Idealization in the Narrow Sense 181
     3.7   Neglect 182
     3.8   Simplification 182
4   Mathematical Simplicity 183
5   Idealization and Reality 184
     5.1   Essentialism 184
     5.2   Idealizations and Empirical Adequacy I 186
     5.3   Idealizations and Empirical Adequacy II 188

6   A Rationale for Abstractions 189
7   Conclusion: Idealization and Symbol 191

9. Symbolizing States and Events in Quantum Mechanics

Carsten Held 193

1   Preliminaries: The Born Rule 194
2   Is the State Vector a Probabilistic System Description?
    The Main Argument 195
3   Two Objections 200
4   Quantum States and Quantum Events 203
5   Quantum Mechanics and the Classical World Picture 207

10. The Semiotics of "Postmodern" Physics

Hans J. Pirner 211

1   Introduction 211
2   Postmodern Fields of Physics 212
3   The Semiotics 217
4   The Semiotics of Postmodern Physics 221
5   Conclusions 227

List of Contributors

Martin Carrier
Fak. für Geschichtswissenschaft
und Philosophie, Abt. Philosophie
Universität Bielefeld
Postfach 100131
33501 Bielefeld, Germany
mcarrier@philosophie.
uni-bielefeld.de

Brigitte Falkenburg
FB 14, Institut für Philosophie
Universität Dortmund
Emil-Figge-Str. 50
44227 Dortmund, Germany
falkenbg@dx1.hrz.
uni-dortmund.de

Massimo Ferrari
Dipartimento di Storia
Università dell'Aquila
Via Roma 33
67100 L'Aquila, Italy
paola@itim.mi.cnr.it

Carsten Held
Philosophisches Seminar I
Albert-Ludwigs-Universität
Werthmannplatz
79085 Freiburg im Breisgau, Germany
cheld@uni-freiburg.de

Andreas Hüttemann
Fak. für Geschichtswissenschaft
und Philosophie, Abt. Philosophie
Universität Bielefeld
Postfach 100131
33501 Bielefeld, Germany
ahuettem@philosophie.
uni-bielefeld.de

Karl-Norbert Ihmig
Fak. für Geschichtswissenschaft
und Philosophie, Abt. Philosophie
Universität Bielefeld
Postfach 100131
33501 Bielefeld, Germany
kihmig@philosophie.
uni-bielefeld.de

Hans-Jürgen Pirner
Institut fü Theoretische Physik
Universität Heidelberg
Philosophenweg 19
69120 Heidelberg, Germany
pir@tphys.uni-heidelberg.de

Enno Rudolph
Philosophisches Seminar
Universitt Luzern
Kasernenplatz 3 / PF 7455
6000 Luzern 7, Switzerland
enno.rudolph@unilu.ch

Ion-Olimpiu Stamatescu
Forschungsstätte
der Evangelischen
Studiengemeinschaft
Schmeilweg 5
69118 Heidelberg

Institut für Theoretische Physisk
Universität Heidelberg
69120 Heidelberg, Germany
stamates@thphys.
uni-heidelberg.de

Part I

Introduction

1. Sources for the History of the Concept of Symbol from Leibniz to Cassirer

Massimo Ferrari

"Ich getraue mir zu behaupten, daß die unauflöslichen Schwierigkeiten, und die wichtigen Streitigkeiten in den Wissenschaften aus
Mangel an Einsicht in die Natur der symbolischen Erkenntnis ent-
standen sind, und daß also die Hebung jener Schwierigkeiten, die
Beilegung jener Streitigkeiten bloß dadurch bewerkstelliget werden
könne, wenn man die Gränzen der symbolischen Erkenntnis in Ansehn-
ung ihres Gebrauchs festsetzte, ihre verschiedenen Arten bestimmte,
und die Symbolik selbst (das Zeichensystem) diesem gemäß einrichte-
te." (Maimom, 1963, p. 265.)

According to Ernst Cassirer, the problem and the concept of the symbolical function may be considered as the core of the philosophical inquiry
(Cassirer, 1985, p. 1). To be sure, many philosophers of the 20th century
agree with Cassirer's point of view. As Alfred North Whitehead remarked,
"symbolism" does not constitute an idle fantasy, but is immanent to human
existence (Whitehead, 1958, p. 61–62). Nevertheless this "immanence" can
be interpreted in different ways. The relationship of the symbols with reality,
the role played by symbols in our mental activity, the various manners of
"constructing the world" on the basis of human symbolisation, are all chal-
lenges to most of the philosophical disciplines.[1] This is particulary true in
the field of epistemology. Questions such as the meaning of symbols within
scientific knowledge, their epistemological status, whether and to what extent
they are constitutive conditions of experience, and eventually the possibility
of considering the "symbolic form" of knowledge as an unavoidable semiotic function of the scientific grasping of reality are problems concerning not
only the "archeology" of the philosophical tradition, but also and firstly the
philosophical tasks as well as our contemporary debates.

This paper aims only at presenting – so to speak – the "prolegomena" to
and some of the main sources of the history of symbolic conceptions of knowledge since Leibniz. In the spirit of Cassirer's systematic–historical reconstruction of the problem of knowledge within the development of the philosophy
and science of modern times, it seems particulary interesting to recognize
(and to compare) two main traditions in the history of the concept of the
symbol: on the one hand, the Leibnizian tradition, which starts with Leibniz's conception of *cognitio symbolica*, and on the other hand, the Kantian

---

[1] For the field of aesthetics, see, for instance, Pochat (1983).

tradition, the core of which is Kant's (transcendental) question about the possibility of experience and, moreover, about the mental functions constituting the possibility of experience itself.

1   Leibniz's "Cognitio Symbolica" and the Theory of Expression

There is no doubt that Leibniz – as Cassirer suggests – occupies a prominent place in the modern history of the concept of the symbol (Cassirer, 1985, p. 3). The key–text that epitomizes Leibniz's point of view on the status of "symbolic knowledge" (*cognitio symbolica*) is the essay *Meditationes de cognitione, veritate et ideis*, which was published in the "Acta eruditorum" in November 1684.[2] Leibniz aims here to locate the *cognitio symbolica*, which he also calls "blind knowledge", in the more general framework of the classification of various kinds of knowledge. First of all Leibniz draws the distinction between *obscure* and *clear* knowledge. We have obscure knowledge every time we are not able to recognize a thing after we have experienced it (for instance, in the case of vague recollections). In contrast, clear knowledge allows us to recognize exactly the things represented, but it can be *indistinct* or *distinct*: it is indistinct when the representation we have is distinguishable from others, but it is difficult to enumerate separately its characteristics (e.g., knowledge of colours, smells and tastes). We have a distinct knowledge when we are able to formulate a nominal definition of the thing, which gives the enumeration of sufficient distinguishing characters, though some elements of the representation the knowledge of which is mediated through sensory perception can remain irreducible, i.e. undefinable; in this case, the knowledge must be called *inadequate*. In opposition, the *adequate* knowledge is based on a complete analysis carried through to the end; the possibility of the *definiendum* is thus guaranteed by means of a real definition, which shows that no contradiction subsists among the characters of the thing. Furthermore, adequate knowledge can be intuitive or symbolic. It seems however very difficult to reach the former, since only knowledge of numbers is similar to this modality of knowledge otherwise exclusively belonging to God; the latter is the typical form of human knowledge, that is, a kind of knowledge for which it is impossible to perform an infinite analysis and to obtain intuition of the whole nature of the objects. "In most cases, however, particularly in a more lengthy analysis," remarks Leibniz, "we do not perceive all at once the whole nature of the objects, but we substitute for the objects themselves defined signs, whose explanation we can omit for the sake of brevity, assuming that we could, if necessary, give one" (Leibniz, 1951, p. 285). He goes on: "I usually designate such knowledge as blind or also as symbolic; we make use of it to a great extent in Algebra, in Arithmetic and nearly everywhere".

---

[2] See Leibniz (1951), p. 283–290. A good survey of the problem of "symbolic knowledge" in the Leibnizian thought is offered by Krämer (1992).

The main result of these reflections on the nature of symbolic knowledge consists thus of the idea that human knowledge cannot avoid the use of symbols and signs. More precisely it may be said that, according to Leibniz, human understanding has no possibility of reaching a kind of knowledge which is at the same time both adequate and intuitive. In contrast, a similar performance is allowed only to divine understanding, and therefore it is undoubtedly correct to affirm that symbolic knowledge represents a "compensation" of the natural bounds of our finite, human reason.[3] However, the Leibnizian conception of the symbolic powers of the mind is grounded on his mathematical and logical research, as it is easy to see from the illuminating *Dialogus de connexione inter res et verba*, which Leibniz wrote in August 1677. Arithmetic is presented here as a kind of *cognitio symbolica*, since it is surely possible to think and reason without words, but it is absolutely impossible to perform any mathematical calculation without "some sign or other". "Ask yourself," says the Leibnizian interlocutor of the dialogue, "whether you can perform any arithmetical calculation without making use of any number-signs" (Leibniz, 1951, p. 18). In Leibniz's opinion the vain attempt to think, speak or calculate without signs, characters or geometrical figures represents the clear proof of the urgency of a "universal characteristic", namely of a systematization and logical treatment of all the characters by means of which the thoughts of the mind and the things of the world are represented (e.g., "the circle drawn on paper is not the true circle; but it is not necessary that it should be, for it suffices to substitute the drawn figure for the circle" – Leibniz, 1951, p. 9). But all this requires that a determined relationship subsists between the signs and the designated things. To be sure, it may be assumed that the order and the connection which are valid for the signs are valid for the things too. But must the sign be similar to the *designatum*? And to what extent must it be arbitrarily assumed? In the *Dialogus* Leibniz gradually drives his interlocutor to the refutation of Hobbesian strict nominalism. "For even though characters are as such arbitrary, there is still in their application and connection something valid which is not arbitrary; namely, a relationship which exists between them and things, and consequently definite relations among all the different characters used to express the same things. And this relationship, this connection is the foundation of truth." Leibniz goes on: "You see that no matter how arbitrarily we choose characters, the results always agree provided we follow a definite order and rule in using the characters" (Leibniz, 1951, p. 10–11). Leibniz also pays attention to the question of the similarity of characters to the designated things, remarking that from the epistemological point of view this similarity has fundamentally no importance. So, for example, the alphabetic letter a is completely dissimilar from the geometrical line which is designated by this letter, and nevertheless the mathematical reasoning works perfectly (Leibniz, 1951, p. 9).

---

[3] See Krämer (1992), p. 227. On the limits of human understanding in the philosophy of 18th century, see Tonelli (1987), p. 45–67, especially p. 46–47.

The foundation of the symbolic knowledge by means of signs and characters is in close connection with the theory of *expression*, which Leibniz first puts forward in the short writing *Quid sit idea* (c. 1678) and to which he refers many times later on.[4] First of all Leibniz proposes the definition of notions such as *expressio* or *exprimere* in the following way: "The means of expression must include structures (*habitudines*) corresponding to the structures of the thing to be expressed."[5] According to Leibniz, there are different kinds of expression: so, for example, the model of a machine expresses the machine itself, a geometrical projection on the plane expresses a three-dimensional figure, a speech expresses opinions and truths and arithmetic or algebraic characters express numbers. Leibniz, in particular, focuses his mind on two main types of expression: on the one hand the expressions which are based upon nature, and on the other hand the expressions which depend on arbitrary conventions.[6] In order to have expressions it suffices that a certain analogy subsists between *expressio* and *res exprimenda*: in the case that this analogy is conceived as a rigorous similarity (*similitudo*), as, for example, when a map expresses the depicted region, we have to do with an expression depending on nature; but in the case of geometrical projection, the relationship between *expressio* and *res exprimenda* belongs to a type of expressions which could be defined more as "functional" than in terms of "similarity".[7] Lastly there are expressions based only upon analogy, and in this case a determinant role is played by the arbitrary stipulation of signs. However the assumption of arbitrary or conventional signs does not prevent at all the fact that also artificial characters maintain some connections to natural expressions. Leibniz's research in the field of natural and artificial languages is the most significant example of this state of affairs.[8]

Generally speaking, it may be said that Leibniz lays the foundations for some modern theories of symbol by means of three main arguments. First of all, Leibniz stresses over and over again the leading role of the *cognitio symbolica*, since human beings can think and know only by having recourse

---

[4] See Leibniz (1951), p. 281–283. However, in the works of Leibniz one can find many passages concerning the theory of expression. See, for instance, Leibniz (1875–1890), vol. II, p. 112, and vol. VI, p. 616–617. One of the most critical expositions of Leibniz's concept of expression is offered by Mugnai (1976), p. 38-61. See also Kulstad (1976), Ghio (1979) and Piro (1990), p. 163-173.

[5] Leibniz (1951), p. 281 (transl. modified). See the original latin passage in Leibniz (1875–1890), Vol. VII, p. 262. "Exprimere aliquam rem dicitur illud, in quo habentur habitudines, quae habitudines rei exprimendae respondent."

[6] See Leibniz (1951), p. 282: "It is also evident that some means of expression have a natural basis and others are at least partly arbitrary, for example, those due to sounds or written characters."

[7] This interpretation is especially asserted by Cassirer (1995), p. 168. See also Mugnai (1976), p. 40.

[8] On this aspect see Mugnai (1976), p. 40–47 and Heinekamp (1972/1975), p. 368–386.

to natural or artifical signs, characters and symbols.[9] Secondly, Leibniz emphazises that the mind is able to express something, and in this context he analyses different kinds of expressions, especially the natural and arbitrary expressions. Thirdly, Leibniz poses the question about the relations subsisting among the signs, on the one hand, and between the signs and the designated things, on the other hand, without assuming however that the signs have to picture the reality. The Leibnizian perspective discloses thus a problem which will be again at the core of modern theories of signs and symbols such as those of Heinrich Hertz, Ludwig Wittgenstein and Moritz Schlick. These three main arguments by Leibniz seem to lead to a more general thesis, according to which signs and symbols play a constitutive role in the process of human knowledge (See Krämer, 1992, p. 225; Heinekamp, 1972/1975, p. 360–361). Here "constitutive" signifies that our discursive knowledge can be achieved only by means of the mediating function of signs, to which we must have recourse in order to speak about things, ideas or state of affairs as well as in order to communicate our thoughts to other human beings. It is undeniable, however, that the genuine significance of "constitutive" belongs only to Kantian critical philosophy, that is, to the reflection on the structure of the mind and on its a priori cognitive abilities. Nevertheless, Leibniz is still fully indebted to the idea of an *armonia praestabilita*: "God, the author of both things and the mind, has endowed our mind with this power to infer from its own internal operations the truth which corresponds perfectly to those of external things." (Leibniz, 1951, p. 282–283).

Excursus 1: Traditions of the Concept of Symbol In the 18th century a conceptual tradition grew, which was a widely influential one for the history of the concept of symbol until Kant's critical philosophy (Lamacchia, 1990, p. 55–97). This quite considerable aspect is connected with the further development of the Leibnizian logic and with its mathematization in the age of the German Enlightement, as is suggested by names such as Gottfried Plouquet, Jakob and Johann Bernoulli and Johann Heinrich Lambert, who were involved in the discussion about both the logical and the metaphysical implications of the Leibnizian universal characteristics (Barone, 1999, p. 60–119; Peckhaus, 1997, p. 64–110). It seems however particularly interesting to stress that *signum* and *symbolum* represent items to which many philosophers of the time paid attention, such as Christian Wolff in his *Psychologia empirica*, Alexander Baumgarten in his *Metaphysica*, Georg Meier in his *Vernunftlehre*, and also Christian August Crusius as well as Lambert. But, interestingly enough, Kant too was well acquainted with this complex tradition, as we can clearly see from his lectures, reflections and published writings. However, as we shall show below, Kant gave to the problem of symbolic knowledge a new systematization, although the Kantian solution was still partially rooted in the German tradition we have recalled (for instance, Crusius and Meier).

As an especially illuminating pattern, it will be sufficient to recall here Lambert's semiotics. In his main work *Neues Organon* (1764), he develops a "theory

---

[9] See Leibniz (1875–1890), vol. VII, p. 31, 191, 204.

of the relation of thoughts to the things" (named precisely semiotics) which firstly deals with the "symbolic knowledge". According to Lambert, symbolic knowledge represents an "unavoidable subsidiary half of thinking", since without conceptual signs we could not avoid becoming overcome by our momentary sensory impressions or aware of already acquired sensations merely in an "obscure and fleeting" way (Lambert, 1965/1968, vol. II, p. 11). Moreover, the fact that knowledge is "completely symbolic" means for Lambert that not only words and signs, but also figures are involved in the human process of knowing (Lambert speaks of "figürliche Erkenntnis"; Lambert, 1965/1968, vol. II, p. 15). It may be also remarked that through these and similar reflections Lambert prefigures just a *tòpos* of the theories of signs and symbols which would be widely influential a century later: especially when he states that the "theory of the things" must be reduced to the "theory of the signs" (Lambert, 1965/1968, vol. II, p. 16). In Lambert's it seems absolutely clear that one has to distinguish between the imitation of a thing (*Nachahmung*), its image (*Bild*) and its sign (*Zeichen*): the sign is by no way a copy (*Abbildung*) of the designated thing (Lambert, 1965/1968, vol. II, p. 16).

Beyond Lambert's semiotics there is another noteworthy aspect. In fact the tradition of the *cognitio symbolica* exerts a quite considerable influence on the rise of the philosophy of language as well. Nobody – as Jürgen Trabant points out – "can call into doubt" the importance of the "great Leibniz" for the philosophy of language (Trabant, 1990, p. 93). To be sure, this is true not only in the case of Herder, whose treatise on the origin of language (1770) was inspired by the reading of Leibniz's *Nouveaux essais*, but also in the case of Condillac, who was well acquainted with Leibnizian philosophy. Furthermore, a similar remark is particularly valid for Wilhelm von Humboldt, whose work may be considered at the same time as the background of Cassirer's view of language as symbolic form.[10] But the role played by Humboldt must also be emphasized for another reason, that is his attempt to relocate the Leibnizian tradition on the ground of Kant's critical philosophy. The Humboldtian synthesis gives thus origin to a new fruitful perspective and seems to assume a paradigmatic significance within the development of the history of the concept of symbol.

2   Kant: Schema and Symbol

The conceptual tradition of the *cognitio symbolica*, which starts with Leibniz, plays an important role also in Kantian thought, although it underlies deep transformations in the framework of critical philosophy.[11] In his so-called "precritical" period, Kant deals with the concept of symbol, as we can see in particular from the following passage of the Dissertation of 1770 offering an illuminating example of the Kantian point of view before the *Critique of*

---

[10] On Humboldt's debt to Leibniz, see Borsche (1981), p. 156ff., Trabant (1990), p. 69ff. and Formigari and De Mauro (1989).

[11] A wide analysis of the role of *cognitio symbolica* in Kant's thought is offered by Lamacchia (1990), p. 70–97.

Pure Reason: "There is (for man) no *intuition* of what belongs to the under-standing, but only a *symbolic cognition*; and thinking is only possibile for us by means of universal concepts in the abstract, not by means of a singular concept in the concrete."[12] But this conception acquires a new significance within the critical philosophy, since the role of symbols is perfectly determined only from the point of view of the transcendental structure of experience, and more precisely in the light of the schematism of the pure concepts of under-standing. The main question for Kant is now represented by the difference between schema and symbol. The schema fulfils the requirement of a third element, "which is homogeneous on the one hand with the category, and on the other hand with the appearance" (A 138/B 177; Engl. trans., p. 181).[13] The transcendental schema is precisely the "mediating representation", which has the property to be at the same time "in one respect [...] *intellectual*, [...] in another [...] *sensible*" (ibid.). For Kant it seems nevertheless unavoidable to distinguish the schema from the mere image. This is very important with regard to the Kantian definition of schema as a procedure which allows us to provide "an image for a concept". This procedure, says Kant, "I entitle the schema of this concept" (A 140/B 179-180; Engl. trans., p. 182). The schema – "this representation of a universal procedure of imagination" – can also be considered as the condition of possibility of the image, although the roots of schematism of pure reason are placed, according to Kant's enigmatic definition, "in the depths of the human soul" which are hard to discover (A 140-141/B 179-180; Engl. trans., p. 182-183).

At this point the more important question is whether the schema may be considered as a candidate for the role of the symbol, which the Leibnizian tradition has trasmitted to Kant. As stressed above, Leibniz (and like Leibniz a lot of post-Kantian views of symbolism we will have to deal with) conceived a kind of isomorphism, but at the same time a kind of dualism too, between the mental sign and the designated (or to the sign co-ordinated) thing. The core of schematism consists instead of the attempt to overcome a similar perspective by means of a "third", pure element of knowledge. The chapter devoted to schematism in the *Critique of Pure Reason* suggests in fact that "the schemata of the pure concepts of understanding are [...] the true and sole conditions under which these concepts obtain relation to objects and so possess significance" (A 146/B 185; Engl. trans., p. 186). The categories acquire their meaning only when the schemata "realize" them and at the same time "restrict" them. It seems in this way that the problem of meaning in general coincides with the problem of the significance of the categories, that is, with the possibility of their schematization on the basis of the "monogram of pure *a priori* imagination". "The categories, therefore, without schemata" –

---

[12] Kant (1992), p. 389. The original latin text affirms: "Intellectualium non datur (homini) Intuitus sed non nisi cognitio symbolica et intellectio nobis tantum licet per conceptus universales in abstracto, non per singularem in concreto."

[13] All the quotations from the *Critique of Pure Reason* are drawn from Kant (1963).

says Kant in the final passage of this quite intricate chapter about schematism
– "are merely functions of the understanding for concepts; and represent no
object. This [objective] meaning they acquire from sensibility, which realises
the understanding in the very process of restricting it" (A 147/B 187; Engl.
trans., p. 187).

Moreover, we can see that, according to Kant, a concept may be symbolized only when there is no possibility to find a schema for it, that is, when it is
impossible to assign to this concept objective reality by means of an intuition
corresponding directly to the concept. Kant believes for this reason that it
would be a systematic fault to take a schema for a symbol or to equate the
one with the other. A similar mistake leads to the "mysticism" that Kant intends to fight in a passage of the *Critique of Practical Reason*: the mysticism
"which makes what served only as a *symbol* into a *schema*, that is, puts under
the application of moral concepts real but not sensible intuitions (of an invis-
ible kingdom of God) and strays into the trascendent" (Kant, 1996, p. 197).
Kant pays attention to this very crucial issue also in the *Preisschrift* in the
Fortschritte der Metaphysik, where he claims that the symbolization of a concept of pure understanding occurs only in the case where objectiv validity
cannot be ascribed to this concept "directly" (*directe*) by a sensible intuition
(this is obviously the schematization's procedure), but only "indirectly" (*indirecte*) by its consequences (Kant, 1981, p. 613). The symbolization is thus
required for all the concepts referred to the intelligible world, i.e., the concepts which cannot be represented through intuitions within the bounds of
possible experience. Therefore the theoretical knowledge of the *Übersinnliche*
(for example, God) has for Kant no ground to subsist: what is allowed in
man's reason with regard to this aspect consists only of a knowledge by analogy. But in this context even the "analogical" knowledge has the significance
of the symbolization of an idea of reason, which deals with the consequences
of an object.[14]

The Kantian concept of symbol seems now clearly defined. Nevertheless
it will be noteworthy to recall here the & 59 of the *Critique of Judgement*
too, which is notoriously devoted to beauty as the symbol of morality. Kant
stresses again that the symbol is able to express the intelligible, and more
particularly it is the determination of aesthetic taste as the medium between

---

[14] Kant (1981), p. 613-614: "Das Symbol einer Idee (oder eines Vernunftbegriffes)
ist eine Vorstellung des Gegenstandes nach der Analogie, d.i. dem gleichen
Verhältnisse zu gewissen Folgen, als dasjenige ist, welches dem Gegenstande an
sich selbst zu seien Folgen beigelegt wird, obgleich die Gegenstände selbst von
ganz verschiedener Art sind, z.B. wenn ich gewisse Produkte der Natur, wie etwa
die organisierten Dinge, Tiere oder Pflanzen, in Verhältnis auf ihre Ursache, mir
wie eine Uhr, im Verhältnis auf den Menschen, als Urheber, vorstellig mache,
nämlich das Verhältnis der Kausalität überhaupt, als Kategorie, in beiden eben
dasselbe, aber das Subjekt dieses Verhältnisses, nach seiner inneren Beschaffenheit mir unbekannt bleibt, jenes also allein, diese aber gar nicht dargestellt
werden kann."

the sensuous stimulus and the intelligible dimension that leads one to conceive the "symbolic" status of beauty. Beauty is thus, according to Kant, an analogical presentation (*exhibitio*) of moral good (Kant, 1973, & 59). However what is of main interest is the reflections that Kant devotes to the concept of symbol in the introduction to this famous paragraph of the *Critique of Judgement*. First of all Kant speaks about the presentation (*Darstellung*) or *Hypotypose*, i.e., about the possibility of schematization or symbolization. By the explanation of the difference between the two concepts we already stressed, Kant takes the opportunity to add a very interesting remark. He says: "Notwithstanding the adoption of the word *symbolic* by modern logicians in a sense opposed to an intuitive mode of representation, it is a wrong use of the word and subversive of its true meaning; for the symbolic is only a mode of any intrinsic connection with the intuition of sensation and is, in fact, divisible into the *schematic* and the *symbolic*. Both are hypotyposes, i.e., presentations (*exhibitiones*), not mere marks. Marks are merely designations of concepts by the aid of accompanying sensible signs devoid of any intrinsic connection with the intuition of the object. Their sole function is to afford a means of reinvoking the concepts according to the imagination's law of association–a purely subjective role. Such marks are either words or visible (algebraic or even mimetic) signs, simply as *expressions* for concepts" (Kant, 1973, & 59).

A comment to this important passage may be subdivided into three aspects:

(a) The "new logicians" make a "wrong use" of the term symbolic. They confuse symbols with characters and thus do not understand the real structure of symbolism. But for even this reason Kant is quite far from the Leibnizian dream of a universal characteristic as well as from the Leibnizian tradition we have recalled above.

(b) The level of symbolic exibition is wholly different from that of mere characters, which are exclusively arbitrary signs accompanying concepts according to the laws of empirical imagination. In addition, since symbolism belongs to an intuitive component of knowledge, having recourse to characters in order to designate the transcendental activity of the mind implies a misleading interpretation of the structure of human knowledge itself. Words, and visible and mimetic signs belong indeed to *Facultas signatrix*, to which Kant refers in his *Anthropologie in pragmatischer Hinsicht*. However, this faculty concerns rather the man as a natural being and citizen of the world (Weltbürger), who needs characters and signs for the scope of communication with others human beings: in this context, Kant says, "the sign (character) accompanies the concept only as its guardian (custos), so that it can reproduce the concept when the occasion arises" (Kant, 1902ff., vol. VII, p. 191). Of course, this pragmatic dimension of language does not exhaust the role of language within Kant's critical inquiry and cannot eclipse the central point represented by the link between language and thought as well as between word and concept; but this question is not to be discusssed here.[15]

(c) Both schemas and symbols are devoted to represent a concept by means of its corresponding intuition, that is, respectively *directe* and *indirecte*. But in the *Critique of Judgement* the role of symbols and analogies (such as that of a constitutional monarchy compared with a body endowed with soul and of a despotic monarchy compared with a mere machine) seems to be more important than it was previously in the *Critique of Pure Reason*. This is surely connected with the role of the regulative ideas of reason as well as of reflective judgement, both of which are involved in the overcoming of the epistemological pattern represented by Newtonian science as a ground for the critical inquiry into the conditions of the possibility of experience. In the new perspective entered by the *Critique of Judgement* Kant is now in the position to conceive the *cognitio symbolica* as a kind of "abbreviated knowledge", which plays a decisive role when the "constitutive" knowledge in the sense of the *Critique of Pure Reason* is not (or still not) possible. Thus, it may be said that in the *Critique of Judgement* symbolization constitutes a very important part of the corpus of knowledge, for we need also symbolized and not only schematized concepts in order to bring the disquieting variety of empirical laws of nature into a system organized by our mind (see Garroni, 1998, p. 98).

There is another aspect that is noteworthy in the context of Kant's discussion about the status of symbol. In the concluding sections of the *Critique of Judgement* Kant introduces the distinction between *intellectus echtypus* and *intellectus archetypus*, placing thereby the roots of the concept of symbol in the systematic background of man as a finite being, who can make no use of intuitive understanding (Kant, 1973, & 77). According to Kant, a similar power of mind has to be conceived only as a regulative idea, whereas the *intellectus echtypus* needs symbols in order to give a sensible representation of those concepts which can only be thought by reason without offering a corresponding intuition *a priori*. It is precisely this Leibnizian conception of symbol as a kind of compensation of the weakness of human reason which throws, in the *Critique of Judgement*, a new and perhaps exiciting light upon Kant's concept of symbolic knowledge.

3   The Rediscovering of the Sign

The renaissance of Leibniz's logic and the renewed interest in his projects of a "universal characteristic" are usually associated with the critical works of Louis Couturat, Bertrand Russell and Ernst Cassirer at the beginnig of the 20th century. But the history of the reception of the Leibnizian logical work is more complicated than the standard view suggests. This circumstance is also important in order to reconstruct the history of the concept of symbol,

---

[15] See especially Capozzi (1987).

which partially coincides with the reception of Leibniz's logic and theory of knowledge. The early rediscovery of the Leibnizian heritage begins indeed in the 1840s, when, in the context of the decline of German speculative idealism, Leibniz was interpreted as an alternative to the development of philosophy after Kant. To be sure, in the intellectual landscape of Germany the rehabilitation of Leibniz's thought was a crucial event, although often neglected by traditional histories of philosophy. Particularly in the field of logic and theory of knowledge, in the middle of the 19th century a kind of "Leibnizian tradition" was established, which also played a considerable role in the development of the conception of symbol.[16]

After the publication of Leibniz's *Opera omnia*, edited by Johann Eduard Erdman in 1840, a widespread interest in the Leibnizian *characteristica universalis* arose as well as in the general philosophy of symbol and sign which was closely related to Leibniz's work on logic. An illuminating example of this renaissance of *esprit leibnizien* can be found in the philosophy of Adolf Trendelenburg, whose attention to Leibniz is also documented in an edition of some of his unpublished writings. In 1857 Trendelenburg delivered a lecture on *Leibnizens Entwurf einer allgemeinen Charakterisktik*, which was devoted to explain on the one hand the importance of Leibniz's project of a *lingua characterica* (according to Trendelenburg's wrong definition), and on the other hand the general significance of the sign as an indispensable instrument of thought. Trendelenburg was convinced that the *characteristica* intended as an *ars inveniendi* and a procedure of calculus was the "more doubtful part" of the whole Leibnizian enterprise (Trendelenburg, 1867, p. 23). But in spite of this the Trendelenburg's lecture offered a good specimen of a general philosophy of sign in agreement with Leibniz's spirit. According to Trendelenburg, the progress of man's thought depends fundamentally on the "designation of the things": the sign represents the means by which our mind can master the reality (Trendelenburg, 1867, p. 1, 4). In speaking and writing, the signs allow the mental representations to be distinguished, which otherwise would be only "streaming representations"; therefore, by means of the signs, the thought can keep itself separate from sensorial impressions and rise in the domain of universality (Trendelenburg, 1867, p. 1). Moreover Trendelenburg stresses that the "presupposition of intellectual sign" constitutes the unavoidable basis for any improvement of the human mind; and in this context, according to Schleiermacher's suggestion, he speaks of a *symbolic* "direction of knowledge" as well as of the "designation's activity (*bezeichnende Thätigkeit*)" which is proper to thinking and knowing (Trendelenburg, 1867, p. 2–3).

Trendelenburg's remarks on the importance of sign and his analyses devoted to Leibniz's great hope of a universal language exerted a certain influence on the German philosophical community. For example Gottlob Frege

---

[16] For the reception of Leibniz's logic in the 19th century, see Peckhaus (1997), esp. p. 130ff.

refers to Leibniz's work in agreement with Trendelenburg, as we can see from a lot of passages of his writings, where Frege particularly discusses the *lingua characterica* (just the same wrong expression we found in Trendelenburg!) and at the same time stresses its too much neglected "difficulties."[17] But it is even more interesting that in his paper *Über die wissenschaftliche Berechtigung einer Begriffschrift* Frege considers the concept of sign nearly in the same attitude we met in the above-quoted lecture of Trendelenburg. Frege holds that the sign is a "great discovery" (*Erfindung*), which allows one to think by means of sensible elements, although becoming free from their "constraint" (Frege, 1980, p. 91–92). Our mental activity comes to its goal only on the basis of signs, i.e., of words as well as of numbers and so on. The world of the intelligible – the world of ideas and concepts – is open for us only via the sensible element of sign. But, on the other hand, by means of a complex of artificial signs it is possible to overcome the ambiguities and the drawbacks of natural languages, which is precisely, of course, the purpose of Frege's "conceptual notation" (*Begriffschrift*) (Frege, 1980, p. 92, 94). "For this ground," exclaims Frege, "nobody can despise the sign!" (Frege, 1980, p. 92).

The case of Frege shows quite well that the logical research in the second half of 19th century finds in Leibniz its mentor or, speaking more precisely, that it believes it possible to place itself, although more *post festum* than from the beginning, in the framework of the Leibnizian tradition (Peckhaus, 1997, p. 299ff). On the other hand, the concepts of symbol and sign became more and more the epistemological focal point of the logical renewal in the 19th century beginning from the rise of algebraic logic in the English scientific community. Small wonder for example that George Boole emphasizes over and over again the role of symbols within his philosophy of logic. He distinguishes between the symbols, which in logical and scientific reasoning are used "with a perfect comprehension of that which renders their use lawful", and the merely arbitrarily posed "characters", "the use of which," says Boole, "is suffered to rest upon authority" (Boole, 1951, p. 10; see also Barone, 2000, p. 83). Boole devotes also a great deal of attention to logic as a discipline of the thought expressed by means of signs or symbols. In his main work Boole defines the investigation of the laws of thought in terms of systematic research on the symbolic process of reasoning. It seems appropriate to quote Boole extensively: "in the process of reasoning, signs stand in the place and fulfil the office of the conceptions and operations of the mind; but that as those conceptions and operations represent things, and the connexions and relations of things, so signs represent things with their connexions and relations; and lastly, that as signs stand in the place of the conceptions of the mind, they are subject to the laws of those conceptions and operations" (Boole, 1950, p. 26). The problem of the meaning of the symbols and of "that which renders

---

[17] See Frege (1879), p. IX–X, and Frege (1979), p. 9 ff. The relationship of Frege to Leibniz is discussed by Patzig (1969), Kluge (1977) and Kluge (1980), p. 231–290.

their use lawful" becomes thus a crucial point, to which Boole goes back also in his manuscripts of the years 1855 and 1856. By reflecting on the nature and office of signs he particularly emphasizes two aspects, i.e., the fact that "signs are arbitrary as concerns their outward form, fixed as concerns their interpretation and their laws", and altogether that "the laws of signs are a visible expression of the formal laws of thought".[18] In this way Boole achieves a general view of sign within the framework of the "laws of thought", which can be considered as the most widespread and shared conceptual background of the philosophy of logic in the late 19th century.

But another interesting example in the field of algebraic logic comes from Ernst Schröder. In 1890, namely in the same year of publication of his *Vorlesungen zur Algebra der Logik*, Schröder delivered a lecture on the thema of sign (*Über das Zeichen*), and, welcoming the suggestions of Leibniz as well as of Trendelenburg, he urged the logicians not to forget the extraordinary importance of signs (see Peckhaus, 1997, p. 286). Schröder points out their relevance also in the introduction to his *Vorlesungen*, which quite literally echoes Trendelenburg's lecture on Leibniz and gives prominence to the Leibnizian *cognitio symbolica*. However, by emphasizing that the mind is able to designate and to symbolize ("bezeichnende" oder "symbolisierende" Thätigkeit – Schröder, 1966, vol. I, p. 38), Schröder does not show at this point any originality: on the contrary, he repeats what had just become commonplace, and therefore he merely strengthens an established view.

Excursus 2: Charles S. Peirce Charles Sanders Peirce also made his contribution to the logical–algebraic research in the second half of the 19th century. However, a typical feature of Peirce's thought is represented at the same time by his aim to reach the goal of a general semiotics, the origins of which are rooted in a revised version of Kant's theory of categories as well as of logic. Interesting enough, Peirce emphasizes in his youthful essay *On a New List of Categories* (1868) that the proper field of logic consists of a systematic research not only into the concepts, but also into all the symbols requested by mental activity (Peirce, 1984, p. 56–57). Furthermore, it is noteworthy that Peirce speaks of signs and symbols according to the Leibnizian definition of *cognitio symbolica*, namely referring to Leibniz's writing *Meditationes de cognitione, veritate et ideis*, which he quotes for the first time in a manuscript of 1866 (Peirce, 1982, p. 355–356). By his attempt to formulate a new list of categories, Peirce outlines a division of the cognitive process into three main categories, placing them between being and substance as the entering and the

---

[18] Boole (1997), p. 130–131: "That signs are arbitrary as to their outward form, is evident from the diversity of languages, the same thing being represented in one language by one combination of letters or sounds and in another language by another. That they are fixed as concerns their interpretation is a truth which is familiary expressed in the rule that the meaning of a word or any other sign must not be ambiguous. Whatever meaning is once given to it, must continue to be associated with it, if language is to be definite as a medium of communication or exact as an instrument of thought".

final terms of this process respectively. A similar triple division (that is: quality, relation and representation) makes possible, firstly, that something is referred to its *ground*; secondly, that something is referred to its correlate; and thirdly, that an interpretant founds the correlation itself as mediating term between the two other terms. This last categorial aspect, that is the *Thirdness*, is decisive, for the infinite process of semiotic interpretation starts from this third level. Peirce determines in 1868 that one has to distinguish between likeness, index or sign and genuine symbol (Peirce, 1984, p. 55–56), although only later on does he wholly elaborate the triple division into icon, index and symbol (see, for instance, Peirce, 1934, p. 50–51). All this is conceived as following: an icon is a *representamen*, which represents something by virtue of a character which it possesses in itself, even if this object does not exist (e.g. a centaur); an index is a *representamen* that fulfills its function by virtue of the existence of something to which it refers; finally, a symbol is a *representamen*, "which fulfills its function regardless of any similarity or analogy with its object and equally regardless of any *factual* connection therewith, but solely and simply because it will be interpreted to be a representamen. Such for example is any general word, sentence, or book." This "symbolic function" is thus comparable with the "function of meaning" (*Bedetungsfunktion*), which constitutes, according to Cassirer, the most important stage of symbolization.

On the other hand, Peirce states in his essay *Questions Concerning Certain Faculties Claimed for Man* (1868) that thinking without signs is impossible. According to him, a thought can be cognized solely by means of signs, and even what *prima facie* seems not to have been thought through signs is really in mind exclusively by virtue of some sign. But Peirce also puts forward another thesis, namely that every thought based upon signs is referred always to other thoughts, for the essence of a sign itself can be characterized even as a continuous reference to something. There is thought – suggests Peirce – only inasmuch as other thoughts exist, that is past thoughts, for thoughts cannot happen in an instant. Every thought is thus a sign of other thoughts (Peirce, 1934, p. 150–151). In this context it must be also emphasized that in this way Peirce outlines a view of knowledge according to which any kind of knowing as a "copy" of the reality must be refused. Object, sign and meaning are indissoluble elements of a unique process of interpretation that cannot rest on any immediate reference of thought to the objects.

4   Hermann von Helmholtz and Heinrich Hertz

It is well known that, in agreement with the main features of the first "back to Kant" moment in Germany, Hermann von Helmholtz's epistemological reflections start from a transformation of the Kantian theory of knowledge into the framework of the psychological and physiological sciences.[19] In his famous lecture *Über das Sehen des Menschen*, delivered at the University of Königsberg in 1855, Helmholtz emphazises that it has become more and more urgent to complete Kant's theory of knowledge through research of Johannes Müller on the specific energy of sensory organs (Helmholtz, 1903, vol. I, p. 99).

---

[19] For a more detailed account of this historical context, see Köhnke (1986), esp. p. 151–157, and Ferrari (1997), p. 10–28.

According to Helmholtz, the more recent scientific discoveries in this field show that Kant's epistemology has been confirmed, for our representations of the external world are in fact conditioned by the sensory organs or, as Helmholtz says, by the "organisation of the mind (*Organisation des Geistes*)" (Helmholtz, 1903, vol. I, p. 115). The focal point of Helmholtz's theory of signs rests to a great extent on this topic. On the one hand, Helmholtz maintains that it is impossible to have knowledge of the essence of external world in itself, and therefore he is firmly persuaded, on the other hand, that signs play a decisive role in mediating between the "organization" of our mind and the reality outside ourselves: "sensory perceptions are only signs denoting the nature of the external world, and the interpretation of these signs has to be learned from experience" (Helmholtz, 1903, vol. I, p. 17). On the basis of the physiological research about the constitution of the sensory organs, it is now possible, according to Helmholtz, to examine the consistency of its results with Kant's transcendental aesthetics, provided that the borderline between the facts of experience and the a priori given forms of intuition must be drawn in a way other than that followed by Kant (Helmholtz, 1903, vol. II, p. 356).

In his *Treatise on Physiological Optics* Helmholtz points out that "the quality of the sensation is in no way identical with the quality of the object by which it is aroused." This is to say that the quality of the sensation depends upon the nervous apparatus which receives the stimulus from the external world; and thus the sensuous quality we can acquire is, according to Helmholtz, "merely a symbol for our imagination" (Helmholtz, 1962, vol I, p. 4). The epistemological thesis deriving from this insight may be considered as one of the most widely influential statements that a scientist of the 19th century had formulated and, for this reason, it is worthy of being extensively quoted: "Our ideas of things cannot be anything but symbols, natural signs for things which we learn how to use in order to regulate our movements and actions. Having learned correctly how to read those symbols, we are enabled by their help to adjust our actions so as to bring about the desired result; that is, so that the expected new sensations will arise" (Helmholtz, 1962, vol. II, p. 19). On the basis of his psychological and physiological research Helmholtz emphasizes also a kind of general theory of signs or a semiotic view, which seem to rest on two main assumptions: on the one hand, the sensation is only a sign denoting the effect of an object, and on the other hand, the use of signs can be characterized essentially through the fact that "the same sign must be always assigned to the same object", without assuming whatever kind of likeness between the sign and the object corresponding to it (Helmholtz, 1903, vol. II, p. 357).

The more general epistemological framework underlying this perspective saw significant changes from Helmholtz's early work to his later philosophy of science (Schiemann, 1998). But this aspect can be omitted inasmuch as the theory of signs we are interested in stressing here is in no way connected, as Moritz Schlick rightly remarked, to a kind of metaphysical realism and rests rather on the assumption that the signs have not to depict the reality in itself, but solely the lawfulness of reality (Schlick, 1978/1979, vol. I, p. 338–339). Beyond every hypothesis on the "causes" of our sensations, the heart of Helmholtz's doctrine lies in the persuasion that a fundamental difference subsists between the signs of the objects and the objects themselves. Thus the signs are not copies or images (*Abbilder*) of the real things, and what is truly important is the fact that between signs and designated things subsists a unique, not ambiguous relation in order to reproduce the law of something existing or happening (Helmholtz, 1977, p. 122). Helmholtz is particularly clear on this subject in his famous lecture of 1878 *The facts in perception*: "Our sensations are indeed effects produced in our organs by external causes; and how such an effect expresses itself naturally depends quite essentially upon the kind of apparatus upon which the effect is produced. Inasmuch as the quality of our sensation gives us a report of what is peculiar to the external influence by which it is excited, it may count as a symbol (*Zeichen*) of it, but not as an image (*Abbild*). For from an image one requires some kind of alikeness with the object of which it is an image [...] But a sign need not have any kind of similarity at all with what it is the sign of. The relation between the two of them is restricted to the fact that like objects exerting an influence under like circumstances evoke like signs, and that therefore unlike signs always correspond to unlike influences" (Helmholtz, 1977, p. 121–122).

The reception of Helmholtz's theory of sign within the theory of knowledge and scientific thought of the late 19th century surely represents a very interesting point. But in this context it will be enough to recall the case of Heinrich Hertz, whose search for a "physics as rigorous science (*strenge Wissenschaft*)" (Fölsing, 1997, p. 500–512) depends to a great extent, from the epistemological point of view, on his continuation and transformation of Helmholtz's doctrine of symbols. In fact, the most quoted and best known passages of Hertz's *Prinzipien der Mechanik* are those concerning the role, in physical knowledge, of our images of reality in order to foresee future experiences. "But the procedure," says Hertz, "which we always make use of in order to derive the future from the past and thus to reach the desired prevision must be the following: we form for ourselves images (*innere Scheinbilder*) or symbols (*Symbole*) of external objects; and the form we give them is such that the necessary consequences of the images in thought are always the images of the necessary consequences in nature of the things pictured" (Hertz, 1963, p. 1). According to Hertz, two presuppositions are necessary to fulfil this epistemological requirement. Firstly, there must be a "certain agreement" between nature and our mind, for only in this way are we able to build models by virtue of which it becomes possible "to anticipate the facts". In this context Hertz defines symbols as a kind of images mirroring the relations constituting the objects (Hertz, 1963, p. 2). Secondly, the symbols depend nonetheless upon the modalities by which the mind can picture the reality ("Abbildungsweise des Geistes") (Hertz, 1963, p. 3); but Hertz also emphasizes that the power of the mind has to be limited by the fact that the rightness of the images is determined through the "force of the things" (Hertz, 1963, p. 48). Hence it may be said, speaking more generally, that we are able to translate the experience into "the symbolic language of the images we have formed for ourselves", and thus to develop our conceptual activity, only by means of signs and according to the necessary structure of the mind (Hertz, 1963, p. 159). Certainly, Hertz stresses over and over again that the goal of knowledge lies in reproducing (*abbilden*) the external reality (Hertz, 1963, p. 160); but all this does not involve the physical knowledge disposing solely of one model of reality: on the contrary, it is possible to elaborate a plurality of models (Hertz, 1963, p. 197–198) (as it is possible to have different symbols for the same thing; Hertz, 1963, p. 2). The choice between different models depends on the simplicity characterizing one model compared to another; thus the simplicity concerns in no way the physical world (it is not a statement a priori about it), but only the systems of symbols we are dealing with (Hertz, 1963, p. 28).

Hertz's theory of symbol and physical knowledge seems thus quite far from Helmholtz's perspective, inasmuch as the psychological and physiological background characterizing the Helmholtzian doctrine is abandoned by Hertz; in the *Prinzipien der Mechanik*, the power of mental activity takes the place of the structure of our nervous apparatus Helmholtz believed to be the decisive aspect for the theory of knowledge.[20] But at this point two questions arise, which both coincide with two different interpretations of Hertz's epistemology. On the one hand, Hertz seems to suggest that the mind creates freely its images or symbols, which have not to be conceived as "copies" of reality, but only as conceptual frameworks which make possible the physical knowledge; in this perspective it is not surprising that a philosopher such as Cassirer attempts to interprete Hertz in the spirit of the Kantian philosophy or – more precisely – of the Neo-Kantianism of the Marburg school. On the other hand, Hertz's epistemology is surely concerned with a kind of isomorphism too which pivots on the necessity that in the science of mechanics the relations subsisting among the objects are equivalent to the relations subsisting among the symbols (Hertz, 1963, p. 9). But once this aspect has been emphasized, it is not too difficult to find in Ludwig Wittgenstein the spiritual heir of Hertz and in some aphorisms of the *Tractatus* the "ontological" background of his epistemology; so, for example, in the statement "we picture facts to ourselves [...] A picture is a model of reality" (*Tractatus*, 2.1 and 2.12).[21] Thus, the history of the concept of symbol goes on and reaches two very different stations, Vienna and Marburg.

---

[20] Regarding the differences between Helmholtz and Hertz, see Dosch (1997), p. 55. See also Heidelberger (1998), esp. p. 21.

[21] All the quotations from Wittegnstein's *Tractatus* are drawn from the English translation (Wittgenstein, 1974).

5   Intermezzo: Hertz and Wittgenstein

In his *Tractatus logico-philosophicus* Ludwig Wittgenstein quotes only few authors: Frege and Russell many times, Kant and Fritz Mauthner both once, but Hertz twice. This reference to Hertz deserves attention, since it is a proof of Wittgenstein's connection with the above-sketched tradition of the concept of symbol. First of all it may be useful to recall some passages of the *Tractatus*. Aphorism 4.04 states: "In a proposition there must be exactly as many distinguishable parts as in the situation that it represents. The two must possess the same logical (mathematical) multiplicity. (Compare Hertz's *Mechanics* on dynamical models.)" These sentences have to be read in the more general framework concerning the relationship between language and the world. In the *Tractatus* the isomorphism characterizing this relationship is widely analysed by means of the concept of "picture" (*Bild*). "A proposition," says Wittgenstein, "states something only in so far as it is a picture" (4.03). Now, according to Wittgenstein, a proposition is "a model of reality as we imagine it" (4.01; see also 4.021), and the proposition constitutes thereby "a description of a state of affairs" (4.023). But all this requires that one clears the ground itself of the formation of propositions as well, and therefore Wittgenstein remarks that "the possibility of propositions is based on the principle that objects have signs as their representatives" (4.0312).[22]

On the one hand it may be said that by these statements Wittgenstein follows essentially the theory of the picture he inherits from Hertz as well as from the tradition of the epistemology of physical knowledge in the 19th century (Majer, 1985, p. 46, 62; see also Janik, 1994/1995). On the other hand, Wittgenstein's agreement with Hertz has nevertheless to be questioned when we consider that the *Tractatus* seems to transform Hertz's dynamic view

---

[22] There is also another issue associated with Hertz's epistemology. In his *Notebooks* and afterward in the *Tractatus* (6.341–6.342), Wittgenstein emphasizes that mechanics represents an attempt to give a description of the world, namely a description which must be able to construct a net of defined rules with which reality may be put in order. Such a method cannot allow anything to be said about the "true" essence of reality, but it nonetheless is characterized by the determination of the form upon which the description of the world depends. "Mechanics," says Wittgenstein, "is one attempt to construct all the propositions that we need for the description of the world according to a simple plan (nach einem Plan)" (Wittgenstein, 1961, p. 36). See also *Tractatus*, 6.343: "Mechanics is an attempt to construct according to a single plan all the true propositions that we need for the description of the world". Moreover in *Tractatus*, 6.361, Wittengenstein writes: "One might say, using Hertz's terminology, that only connections that are *subject to law* are *thinkable*". All these aphorisms constitute an attempt to develop Hertz's insights about the construction of symbolic systems in order to describe the physical reality. However, it should be discussed whether Wittgenstein agrees with Hertz's point of view or he intends to revise it, which implies a change of perspective precisely about the role of the mind.

about the method of cognition into a static view about language, which appears closer to Helmholtz's conception of a fixed alphabet of thought.[23] But the two interpretations do not have to be understood as mutually exclusive, at least since we have seen that Hertz himself fluctuates between insistence on the creative powers of mind and what Wittgenstein defines as "the logic of depiction" (4.015). There is no doubt, however, that the meaningful propositions of language are formed, according to Wittgenstein, on the basis of the "internal relation of depicting that holds between language and the world".[24] And once we consider that each kind of language depends upon a such isomorphism as well as upon the functional relations among the signs representing a state of affairs, it seems correct to suppose that beyond Hertz's figure it is the work of the great Leibniz which emerges from the *Tractatus*, namely of the Leibniz we encountered by reading *Quid sit idea*. In the fact, some passages of the *Tractatus* may be interpreted in this way, for example, "The pictorial relationship consists of the correlations of the picture's elements with things" (2.1514) or "What any picture, of whatever form, must have in common with reality in order to be able to depict it – correctly or incorrectly – in any way at all, is logical form, i.e. the form of reality" (2.18). To be sure, at first glance it is surprising to find a similar affinity between Wittgenstein and Leibniz. But we shall learn below from Cassirer that the theory of expression outlined by Leibniz in *Quid sit idea* seems to agree with the thesis put forward by Hertz in his *Prinzipien der Mechanik*. From this point of view it is thus possible to assume that via Hertz the *Tractatus* reproposes, although not intentionally, an essential feature of the Leibnizian way of thinking.

Hence, even Wittgenstein's genius has to be located within the tradition of the concept of symbol. But he belongs to this tradition for another reason as well. In the *Tractatus* and in his *Notebooks* (1914–1916) Wittgenstein pays attention over and over again to the concept of symbol. According to Wittgenstein, it is necessary to distinguish between sign and symbol for the reason that the sign represents only the sensuous part of the symbol. "A sign," states Wittgenstein, "is what can be perceived of a symbol" (3.32). So it corresponds to a determinate logical connection of signs, "a determinate logical combination of their meanings" (4.466); but in the case where the meanings do not exist, the relations among signs also have no meaning, that is "they are not essential to the symbol" (as in the case of tautology or contradiction) (4.4661). Wittgenstein intends to stress that in order to recognize a symbol by its sign "we must observe how it is used within a sense" (3.326); therefore a symbol is only meaningfully used when it expresses a sense or a meaning (see 4.5). Nevertheless, in Wittgenstein's opinion it is particularly important to make it clear that meaning does not depend on individual sym-

---

[23] See Majer (1998), p. 230, which rectifies his above-quoted opinion.

[24] See 4.014: "A gramophone record, the musical idea, the written notes, and the sound-waves, all stand to one another in the same internal relation of depicting that holds between language and the world."

bols, but rather on the system through which symbols are organized.[25] A similar idea is stressed in Wittgenstein's *Notes on Logic* (September 1913), where he states the relative independence of individual symbols expressed by signs from the validity of the symbolic system of which they are part: "Man possesses an innate capacity for constructing symbols with which some sense can be expressed without having the slightest idea what each word signifies. The best example of this is mathematics, for man has until recently used the symbols for numbers without knowing what they signify or that they signify nothing" (Wittgenstein, 1961, p. 95–96). This is also important for another reason, since what is more relevant in a proposition is the circumstance, according to Wittgenstein, that the depicted fact is possible, not that it is actually given: "The propositional sign guarantees the possibility of the fact which it presents (not, that this fact is actually the case)" (Wittgenstein, 1961, p. 27). Our pictures of reality are thereby related to the possibility of the states of affairs they depict, but without supposing whatever similarity of the signs to the designated things and, moreover, without supposing that an individual symbol can be understood without referring it to the whole to which it belongs. The heritage of Hertz and, indirectly, of Leibniz seems thus to constitute the background of these reflections, although the status of logical form of language, which cannot be demonstrated but only "shown", represents the typical Wittgensteinian problem and marks his distance from the tradition.

6   From Moritz Schlick to Ernst Cassirer

For the members of the Vienna Circle the relationship of language to the world also resides at the core of the philosophical–epistemological inquiry. To be sure, Wittgenstein's influence on the development of logical empiricism in the 1920s is a point to be discussed in a perspective quite different from that suggested by the received view. But in the context of the story we are reconstructing, the very important circumstance is that also a prominent spokesman of the "scientific worldview" such as Moritz Schlick shares the symbolic conception of knowledge supported by both Helmholtz and Hertz. Through these leading figures of the scientific culture of the late 19th century, and especially through Hertz's theory of symbol, the relationship of Schlick with Wittgenstein becomes more and more clear. In other words, by considering the role of signs and symbols in both of their theories of knowledge and language, it seems doutblessly plausible to maintain that a kind of "preestablished harmony" subsists between Schlick and the *Tractatus* (Haller, 1993, p. 114).

Schlick showed appreciation over and over again for Helmholtz's contributions to epistemology (see, above all, Schlick, 1978/1979, vol. I, p. 335–342),

---

[25] See 5.555: "But where there is a system by which we can create symbols, the system is what is important for logic and not the individual symbols."

and his admiration was so deep that he edited in 1921 a collection of the main of Helmholtz's writings on this subject (Helmholtz, 1977). His notes and comments to Helmholtz's lecture *The facts in perception* reveal how he agreed with the core of the Helmholtzian theory of signs, that is, with the idea that the essence of all knowledge consists of forming "such an image of what is lawlike in the actual, with the help of a sign." Thus it is exclusively on the basis of this method that it becomes possible to fulfill the scope of human knowledge (Helmholtz, 1977, p. 166, note 15). By asserting this thesis, however, Schlick abandons at the same time an essential feature of Helmholtz's doctrine, that is, the psychological and physiological background on which the function of signs, according to his perspective, is necessarily rooted; in this context the role of intuition itself, especially with regard to the development of non-Euclidean geometries, must clearly be overcome (see Friedman, 1997, p. 38–41). For this reason, Schlick's point of view seems to have a more significant debt to Hertz's *Prinzipien der Mechanik*, whose conceptions about symbols is a central one for Schlick and pervades throughout his early philosophy, particularly his great work on the *General theory of knowledge* (first edition 1918, second and revised edition 1925).

Schlick maintains a kind of semiotic conception, according to which concepts are essentially signs, namely signs or symbols "for all those objects whose properties include the various defining characteristics of that concept" (Schlick, 1974, p. 20). Hence the concepts are something "unreal", which have to be represented through "some mental reality (*etwas psychisch Reales*)". So the concept in itself does not exist at all: there is not, precisely speaking, any concept as such, but merely a "conceptual function" (Schlick, 1974, p. 21–22). This function lies in coordinating signs or symbols to reality, which cannot be ever known by intuition, but only grasped on the ground of its designation. "Epistemologically, the import of the conceptual function consists precisely in *designating*. Here, however, to designate means nothing more than to *coordinate* (*Zuordnen*). To say that objects fall under a certain concept is to say that we have coordinated them with this concept."[26]

Hence, according to Schlick, knowledge consists of a collection as well as a system of signs, which stands in a wholly determined and unique relation to reality; the signs, on the other hand, are nothing but what we usually call concepts (in the first edition of the *General Theory of Knowledge*, Schlick speaks even of "fictions") (Schlick, 1918, p. 23). The structure of our knowledge requires therefore that concepts have to be signs representing objects, whereas judgements be signs representing the relations subsisting among the objects. Mastering the world by means of thought means for Schlick that we must master even the signs which are coordinated to the world or to the facts, and truth consists only of this univocal relationship. "Writing or calculating or speaking, like numbering," says Schlick, "is working with symbols, and so

---

[26] Schlick (1974), p. 23 (transl. slightly modified). On the coordination of signs to objects as the keystone of knowledge, see also Carnap (1979), p. 19.

is thinking. To say that in thought we are masters of the world is to say that we are masters of the thoughts and judgments that serve us as signs for all the objects and facts of the world. We carry out these coordinations all the time in ordinary life. But if they are to reach their goal of making symbols authentic representatives of that which is designated, the coordinations must satisfy one essential condition: they must be *unique*, they must tell us exactly which object belongs to a particular sign [...] Now this also holds which regard to the correspondence of judgments with facts. And a judgement that *uniquely designates* a set of facts is called *true*" (Schlick, 1974, p. 59–60).

It would be interesting to compare Schlick's thesis with Wittgenstein's *Tractatus*, in order to discuss both the "pre-established harmony" we have suggested above and the unconditioned admiration, which Schlick showed later reading Wittengstein's masterpiece. It seems however undeniable that Schlick also maintains a kind of Wittegensteinian isomorphism, the background of which can be (at least partially) identified with Hertz's theory of *Bilder*. But precisely for this reason, there is also another context in which it may be useful to locate Schlick's epistemological perspective. In fact, it is the philosopher of the symbolic forms who tackles extensively Schlick's view of knowledge as coordination, stressing in particular that in the *General Theory of Knowledge* two elements seem to be involved and related to each other: "on the one hand the fictionalism of concept," remarks Cassirer, "and on the other hand the realism in the theory of reality" (Cassirer, 1985, p. 126). Cassirer opposes Schlick's point of view proposing a transcendental theory of experience and knowledge, which is characterized by the aim of overcoming the mere identification of the role of symbols as well as of symbolic function with the designating activity of mind. According to Cassirer, the sign has to be understood as the expression of a meaning (*Bedeutung*) and depends thereafter on the general conditions of the possibility of mental reference to reality as well as of mentally making sense of it. Thus, Cassirer's objection concerns the circumstance that Schlick is unable to comprehend that a thought consisting of mere fictions, i.e., concepts exclusively conceived as signs, cannot shape or constitute reality (Cassirer, 1985, p. 138). The question about the conditions of possibility of experience must be transformed from the standpoint of Cassirer's philosophy of symbol into a question about the conditions of possibility of meaning. In Cassirer's opinion a sign is really a sign only when it is closely associated with a sense (*Sinn*), "to which the sign tends and by means of which it becomes 'signifying'" (Cassirer, 1985, p. 136). The problem of such an attribution of a sense to the sign, that is, the problem of finding a principle by virtue of which "a sensuous element is able to become representative of a 'sense'", constitutes "the general problem of meaning (*das allgemeine Bedeutungsproblem*)", which goes essentially beyond the merely negative characterization of the sign as a conventional or arbitrary element. The genuine task of a philosophy of symbols and signs is thus to provide a general theory of mental expression (*geistiger Ausdruck*), and it is only within this framework that scientific knowledge can reach the status of a symbolic form.[27]

7 Knowledge as Symbolic Form

We go back in this way to the work of Cassirer, which represents the starting point of our story. Cassirer's philosophy can be considered as a synthesis of both of the Leibnizian and Kantian central themes we have followed in their developments through two centuries of philosophical and scientific debates. To be sure, the Neo-Kantianism of the Marburg school which constitutes the framework of Cassirer's epistemological work has to be characterized also as a kind of "contamination" with the Leibnizian heritage, especially as regards precisely the symbolic status of knowledge.

It is not by accident that in the introduction to the first volume of his *Philosophy of Symbolic Forms* Cassirer emphasizes the extraordinary importance of Leibniz's "universal characteristics". The greatest merit of the Leibnizian project lies in the exact establishment of the "function of symbolism (*Zeichengebung*)", according to which "the logic of things, i.e., of the material concepts and relations on which the structure of a science rests, cannot be separated from the logic of the signs." In Cassirer's opinion, thus, the sign is "no mere accidental cloak of the idea, but its necessary and essential organ", namely the instrument by virtue of which the content of the thought may be expressed: "the conceptual definition of a content goes hand in hand with its stabilization in some characteristic sign" (Cassirer, 1953b, p. 85–86). Also in the third volume of Cassirer's main work, this crucial point is stressed very well. Thanks to Leibniz – says Cassirer – "at one stroke the concept of the symbol has become the actual focus of the intellectual world" (Cassirer, 1957, p. 46). Of course, Cassirer means particularly to recall to the mind the scientific performance which Leibniz has accomplished in the field of mathematics and symbolic logic by means of his universal characteristics, namely the great transformation of the relation between logic and mathematics that has deeply influenced the development of scientific thought in the late 19th and in the 20th century (Cassirer, 1985, p. 3).

In his youthful book on *Leibniz' System* Cassirer points out that the concept of symbol is the core of the Leibnizian aesthetics, since art is conceived as "pure symbolization (*Symbolik*) of feeling". Such a remark is interesting for us, because in this context Cassirer refers both to the theory of expression developed in *Quid sit idea* and to section 61 of the *Monadology*, where Leibniz makes the statement: "Les composés symbolisent avec les simples"

---

[27] Interestingly enough, Schlick later recognized, in his lectures on *Form and content* (1929), the central role of the expression and its difference from the mere representation: the former concerns the structure within which the symbols (of language, above all) acquire their meaning, whereas the latter deals with the signs we use to designate the things (Schlick, 1938, p. 151–174).

(Cassirer, (1902), p. 464–466). Interestingly enough, two years later Cassirer mentions this last passage in his edition of Leibniz's writings, remarking that the relation between the sphere of monads and that of sensible phenomena can be defined as a relation which brings to expression this state of affairs "in symbolic form" (Leibniz, 1966, vol. I, p. 173, note 114). For the first time, thus, the most famous term of the Cassirerian philosophical vocabulary appears, to be sure not accidentally, in the framework of a Leibnizian problem. But what is, more generally, the meaning of these early "variations" on themes such as the symbol as well as symbolic form? The answer to this question can be found in the chapter that Cassirer devotes to Leibniz in the *Erkenntnisproblem* (1907), where he stresses quite clearly the unavoidable role of symbols within scientific knowledge and – what is for us particularly significant – at the same time he quotes many passages from the Leibnizian *Quid sit idea*. Cassirer suggests that this text can support a *functional* interpretation of Leibniz's doctrine of expression, according to which signs and symbols express the content of ideas and the relations subsisting among ideas without leading, however, to the usual conception of knowledge as a "copy" of the reality outside ourselves (Cassirer, 1995, vol. II, p. 160). In this way the Neo-Kantian Cassirer shows how his epistemological view of the status of symbolic knowledge is deeply connected with the Leibnizian tradition we have followed through two centuries.[28]

There is however another aspect, which has remarkable importance in the context of our story. In the chapter on Leibniz in *Erkenntnisproblem*, Cassirer links the Leibnizian theory of expression to Hertz's epistemological view of symbols. In Cassirer's opinion there is a surprising kinship between both the perspectives, since symbols are conceived by Hertz in the same sense as by Leibniz, namely as ideal constructions of the mind having no similarity with the reality they must represent (Cassirer, 1995, vol. II, p. 168, note 1). To be sure, Cassirer overlooks the isomorphism that is shared both by Leibniz and Hertz, and he fixs his attention on the functionalism we have pointed out above as a leitmotif of his interpretation of Leibniz. As regards Hertz, however, Cassirer is perhaps one of the few philosophers of the 20th century that has grasped his epistemological relevance as forerunner of a "new phase in the methodology of physics", which lies on the theory of symbolization as pure "*activity of thought*" beyond the bounds of the senses. This activity represents, according to Cassirer, a quite complex process of the mind, which aims to determine the possibility of a "pure natural science" in in the Kantian sense. And therefore nothing is more detached from Hertz than the empiristic point of view asserted by philosophers of science such as, for instance, Ernst Mach: "The principles of theoretical physics," says Cassirer, "are thus, according to Hertz, only images anticipating (*Vorbilder*) possible experiences, whereas they are, from Mach's point of view, reproductions (*Nachbilder*) and pictures (*Abbilder*) of real experiences" (Cassirer, 1991, p. 112–113). In this

---

[28] On this issue see also Ferrari (1996), p. 171–189.

way Cassirer attempts clearly to locate Hertz within the tradition of "pure thought" as foundation of every form of reality developed by the Marburg school; but a similar perspective has its advantages too, since Cassirer is doubtlessly right when he states that the main epistemological role of symbols in the sense Hertz elaborated them consists of introducing concepts such as matter, force, atom, aether and so on, which represent the basis for the construction of a physical world "ordered by law", in spite of the fact that they do not correspond to any sensible perception (Cassirer, 1953b, p. 85).

From Leibniz to Hertz, interpreting both of their theories of the symbol in the framework of a Neo-Kantian theory of knowledge seems to constitute, generally speaking, the essential feature of Cassirer's philosophical adventure in the wide field of the *cognitio symbolica*. In 1910, thanks to the work of a great philosopher of science, Pierre Duhem[29], Cassirer is in fact in the position to formulate what has to be the proper maxim of knowledge as "symbolic form": "the concepts of science are no more imitations of existing things, but only symbols ordering and connecting the reality in a functional way" (Cassirer, 1995, vol. I, p. 3). It belongs also to the essential nature of the symbol that the objectivity we ascribe to it does not depend in any way on the objective character of the reality, but only on the conditions of validity of the symbol itself. "The objects of physics are thus, in their connection according to law, not so much "signs of something objective," remarks Cassirer, "as rather objective signs, that satisfy certain conceptual conditions and demands" (Cassirer, 1953a, p. 305). This statement, which Cassirer makes referring to Helmholtz's theory of signs, shows indeed the great extent to which the conception of knowledge as a "symbolic form" rests on some thesis drawn from Hertz's *Prinzipien der Mechanik*. But Cassirer is able to establish the symbolic conception of knowledge only provided that both the realism of Hertz and the schematism of Kant's pure categories are questioned, for it is only "pure" thought that forms meaningful symbols or signs. In order to reach this goal, we have to deal neither with the reality beyond the limits of the constitutive powers of the mind, nor with the schematized concepts of understanding in the Kantian sense, for Cassirer denies – according to the fundamental view of the Marburg school – that sensible intuition represents an autonomous source of knowledge. In fact, the categories of scientific thought are always meaningful inasmuch as they are expression of mental meanings through sensible signs, but they do not need, for this reason, to be "realized" and "restricted" by schematas as mediating representations between pure thought and pure intuition. "'Symbolic form' means every mental energy (*Energie des Geistes*)," so states Cassirer's famous definition, "by means of which a mental content is linked to a sensible sign and is closely attached

---

[29] Duhem (1981), p. 298: "Les faits d'expérience, pris dans leur brutalité native, ne sauraient servir au raisonnement mathématique; pour alimenter ce raisonnement ils doivent étre transformés et mis sous forme symbolique". For the relationship of Cassirer with Duhem's epistemology see Ferrari (1995).

to it" (Cassirer, 1983, p. 175). This general definition of "symbolic form" is valid too for scientific knowledge, since all knowledge – in a quite different way from Kant – is symbolic and only as symbolic is it really possible.

In the *Philosophy of Symbolic Forms*, Cassirer gives a clear account of this perspective, especially when he put forwards that the function of symbolism (*Zeichengebung*) coincides throughout with that of signifying. The mental meaning is gained only by making use of a sign or a symbol, and therefore the sign has to be conceived as "the first stage and the first demonstration of objectivity", as the tool by virtue of which the universal and the particular, the intelligible and the sensible are related to each other (Cassirer, 1953b, p. 86, 89, 105–106). In modern scientific thought Leibniz is the first, according to Cassirer, to understand that this aspect represents precisely the keystone of human reason; but Leibniz is also the philosopher whose insights into the nature of mathematical and logical thinking are still the most influential on epistemological debates of our times, in many respects even more influential than those of Kant (see, for instance, Cassirer, 1957, p. 363–364).

Once again a philosophy of knowledge as a symbolic form appears thus indebted to the Leibnizian heritage, although the constitutive role of symbols presupposes for Cassirer an inquiry into the structure of the mind based on the Kantian analysis of our a priori cognitive powers. Obviously, the consistence of this synthesis can and must be discussed, although in another context and within a more theoretical perspective. Nevertheless, Cassirer seems to represent an excellent standpoint from which the history we have followed up to his conception of knowledge as a symbolic form may be considered; perhaps a similar view is also the key for a close understanding of the origins and further developments of that "Proteus" which the philosophical tradition has named "symbol" (Cassirer, 1985, p. 1).

References

D. Baird, R.I.G. Hughes, A. Nordmann (Eds.) (1998), Heinrich Hertz: Classical Physicist, Modern Philosopher, Dordrecht/Boston/London: Kluwer Academic Publishers.

F. Barone (1999), Logica formale e logica trascendentale, I, Da Leibniz a Kant (1957), Milano: Unicopli.

F. Barone (2000), Logica formale e logica trascendentale, II, L'algebra della logica (1965), Milano: Unicopli.

G. Boole (1950), An Investigation of the Laws of Thought on which are founded the mathematical theories of Logic and Probabilities (1854), New York: Dover Publications.

G. Boole (1951), The Mathematical Analysis of Logic. Being an Essay towards a Calculus of Deductive Reasoning (1847), Oxford: Basil Blackwell.

G. Boole (1997), Selected Manuscripts on Logic and its Philosophy, edited by I. Grattan-Guinness and G. Bornet, Basel-Boston-Berlin: Birkhäuser Verlag.

T. Borsche (1981), Sprachansichten. Der Begriff der menschlichen Rede in der Sprachphilosophie Wilhelm von Humboldts, Stuttgart: Klet-Cotta.

M. Buzzetti and M. Ferriani (Eds.) (1987), Speculative Grammar, Universal Grammar and Philosophical Analysis of Language, Amsterdam/Philadelphia: John Benjamins Publishing Company.

M. Capozzi (1987), Kant on Logic, Language and Thought, in: M. Buzzetti and M. Ferriani (Eds.), pp. 97–147.

R. Carnap (1979), Der logische Aufbau der Welt (1928), Frankfurt am Main/Berlin/Wien: Ullstein.

E. Cassirer (1902), Leibniz' System in seinen wissenschaftlichen Grundlagen, Marburg: Elwert.

E. Cassirer (1953a), Substance and Function and Einstein's Theory of Relativity (1910–1923), transl. by W. Curtis Swabey and M. Collins, New York: Dover Publications.

E. Cassirer (1953b), The Philosophy of Symbolic Forms, vol. I: Language (1923), New Haven: Yale University Press.

E. Cassirer (1957), The Philosophy of Symbolic Forms, vol. III: The Phenomenology of Knowledge (1929), New Haven: Yale University Press.

E. Cassirer (1983), Wesen und Wirkung des Symbolbegriffs, Darmstadt: Wissenschaftliche Buchgesellschaft.

E. Cassirer (1985), Symbol, Technik, Sprache. Aufsätze aus den Jahren 1927–1933, edited by E.W. Orth and J.M. Krois, Hamburg: Meiner.

E. Cassirer (1991), Das Erkenntnisproblem in der Philosophie und Wissenschaft der neueren Zeit, vol. IV: Von Hegels Tod bis zur Gegenwart (1957), Darmstadt: Wissenschaftliche Buchgesellschaft.

E. Cassirer (1995), Das Erkenntnisproblem in der Philosophie und Wissenschaft der neueren Zeit, second revised edition, (1910–1911), 2 vols., Darmstadt: Wissenschaftliche Buchgesellschaft.

H.-J. Dahms (Ed.) (1985), Philosophie, Wissenschaft, Aufklärung. Beiträge zur Geschichte und Wirkung des Wiener Kreises, Berlin-New York: De Gruyter.

T. De Mauro and L. Formigari (Eds.) (1989), Leibniz, Humboldt and the Origins of Comparative Linguistic, Amsterdam/Philadelphia: John Benjamins Publishing Company.

H.G. Dosch (1997), The Concept of Sign and Symbol in the Work of Hermann Helmholtz and Heinrich Hertz, Etudes de Lettres 1-2: 74–61.

P. Duhem (1981), La théorie physique. Son object et sa structure (1906), Paris: Vrin.

M. Ferrari (1995), Ernst Cassirer und Pierre Duhem, in: E. Rudolph and B.O. Küppers (Eds.) (1995), pp. 177–196.

M. Ferrari (1996), Ernst Cassirer. Dalla scuola di Marburgo alla filosofia della cultura, Firenze: Olschki.

M. Ferrari (1997), Introduzione a Il neocriticismo, Roma- Bari: Laterza.

A. Fölsing (1997), Heinrich Hertz. Eine Biographie, Hamburg: Hoffmann und Campe.

G. Frege (1879), Begriffschrift, eine der aritmetischen nachgebildete Formelsprache des reinen Denkens, Halle: Nebert.

G. Frege (1979), Posthumous Writings, edited by H. Hermes, F. Kambartel, F. Kaulbach, Oxford: Blackwell.

G. Frege (1980), Funktion, Begriff, Bedeutung, edited by G. Patzig, Göttingen: Vandenhoeck and Ruprecht.

M. Friedman (1997), Helmholtz's Zeichentheorie and Schlick's Allgemeine Erkennt-nislehre: Early Logical Empiricism and Its Nineteenth-Century Background, in: Philosophical Topics XXV 19–50.

E. Garroni (1998), Estetica ed epistemologia. Riflessioni sulla "Critica del Giudizio", Milano: Edizioni Unicopli.

M. Ghio (1979), Il concetto di espressione in Leibniz e nella tradizione platonico-cristiana, Torino: Edizioni di "Filosofia".

R. Haller (1993), Neopositivismus. Eine historische Einführung in die Philosophie des Wiener Kreises, Darmstadt: Wissenschaftliche Buchgesellschaft.

M. Heidelberger, From Helmholtz's Philosophy of Science to Hertz's Picture-Theory, in: D. Baird-R.I.G. Hughes-A. Nordmann (Eds.) (1998), pp. 9–24.

A. Heinekamp (1988), Natürliche Sprache und allgemeine Charakteristik bei Leib-niz (1972/1975), in: Leibniz Logik und Metaphysik, A. Heinekamp-F. Schuppe (Eds.), (Wissenschaftliche Buchgesellschaft, Darmstadt) pp. 349–386.

A. Heinekamp and F. Schuppe (Eds.) (1988), Leibniz' Logik und Metaphysik, Darm-stadt: Wissenschaftliche Buchgesellschaft.

H. Hertz (1963), Die Prinzipien der Mechanik in neuem Zusammenhange dargestellt (1894), Darmstadt: Wissenschaftliche Buchgesellschaft.

A. Janik (1994/1995), How Did Hertz influence Wittgenstein's Philosophical De-velopment? Grazer Philosophische Studien 49: 19-47.

I. Kant (1902ff.), Kant's Gesammelte Schriften. Akademie Ausgabe, vols. I–XII, Berlin: Reimer.

I. Kant (1963), Critique of Pure Reason, transl. by N. Kemp Smith, London: Macmillan.

I. Kant (1973), The Critique of Judgement, transl. by J. Creed Meredith, Oxford: Clarendon Press.

I. Kant (1981), Schriften zur Metaphysik und Logik, edited by. W. Weischedel, Frankfurt am Main: Suhrkamp (Werkausgabe, vol. VI).

I. Kant (1992), Theoretical Philosophy 1755–1770, transl. and edited by D. Walford in collaboration with R. Meerbote, Cambridge: Cambridge University Press.

I. Kant (1996), Practical Philosophy, transl. by M.J. Gregor, general introduction by A. Wood, Cambridge: Cambridge University Press.

E.-W. Kluge (1977), Frege, Leibniz et alii, Studia Leibnitiana IX: 266-274.

E.-W. Kluge (1980), The Metaphysics of Gottlob Frege. An Essay in Ontological reconstruction, The Hague: Martinus Nijhoff.

K.Ch. Köhnke (1986), Entstehung und Aufstieg des Neukantianismus. Die deutsche Universitätsphilosophie zwischen Idealismus und Positivismus, Frankfurt am Main: Suhrkamp.

S. Krämer (1992), Symbolische Erkenntnis bei Leibniz Z. Philos. Forsch. XLVI: 224-237.

M.A. Kulstad (1976), Leibniz Conception of Expression, Stud. Leibnitiana IX: 56-76.

A. Lamacchia (1990), Percorsi kantiani, Bari: Levante Editori.

J.H. Lambert (1965/1968), Philosophischen Schriften, ed. by H.-W. Arndt, 10 vols., Hildesheim: Olms.

G.W. Leibniz (1875–1890), Philosophische Schriften, edited by C.I. Gerhardt, 7 vols., Berlin: Weidemann.

G.W. Leibniz (1951), Selections, edited by Ph.P. Wiener, New York: Charles Scribner's Sons.

G.W. Leibniz (1966), Hauptschriften zur Grundlegung der Philosophie (1904/1906), edited by E. Cassirer, 2 vols., Hamburg: Meiner.

S. Maimon (1963), Versuch über die transzendentale Philosophie mit einem Anhang über symbolische Erkenntnis und Anmerkungen (1790), Darmstadt: Wissenschaftliche Buchgesellschaft.

U. Majer (1985), Hertz, Wittgenstein und der Wiener Kreis, in: H.-J. Dahms (editor) (1985), pp. 40–66.

U. Majer (1998), Heinrich Hertz's Picture-Conception of Theories: Its Elaboration by Hilbert, Weyl, and Ramsey, in: D. Baird-R.I.G. Hughes-A. Nordmann (Eds.) (1998), pp. 225–242.

M. Mugnai (1976), Astrazione e réaltà. Saggio su Leibniz, Milan: Feltrinelli.

G. Patzig (1969), Leibniz, Frege und die sogennante "lingua characterica universalis", in: Studia Leibnitiana Supplementa, vol. III, Wiesbaden: Steiner Verlag, pp. 103– 112.

V. Peckhaus (1997), Logik, Mathesis universalis und allgemeine Wissenschaft. Leibniz und die Wiederentdeckung der formalen Logik im 19. Jahrhundert, Berlin: Akademie Verlag.

C.S. Peirce (1934), Collected Papers, edited by Ch. Hartshorne and P. Weiss, vol. V: Pragmatism and Pragmaticism, Cambridge: Harvard University Press.

C.S. Peirce (1982), Writings. A Chronological edition, edited by M.H. Fish, vol. I: 1857–1866, Bloomington: Indiana University Press.

C.S. Peirce (1984), Writings. A Chronological edition, edited by M.H. Fish, vol. II: 1867–1871, Bloomington: Indiana University Press.

F. Piro (1990), Varietas identitate compensata. Studio sulla formazione della metafisica di Leibniz, Napoli: Bibliopolis.

G. Pochat (1983), Der Symbolbegriff in der Ästhetik und Kunstwissenschaft, Köln: DuMont Buchverlag.

E. Rudolph and B.O. Küppers (editors) (1995), Kulturkritik nach Ernst Cassirer, Hamburg: Meiner.

G. Schiemann (1998), The Loss of World in the Image. Origin and Development of the Concept of Image in the Thought of Hermann von Helmholtz and Heinrich Hertz, in: D. Baird-R.I.G. Hughes-A. Nordmann (editors) (1998), pp. 25–38.

M. Schlick (1918), Allgemeine Erkenntnislehre, Berlin: Springer.

M. Schlick (1974), General Theory of Knowledge, transl. by A. Blumenberg, Vienna: Springer.

M. Schlick (1938) Gesammelte Aufsätze 1926-1936, hrsg. von F. Waismann, Wien, Gerold; english translation: Philosophical Papers, edited by H. Mulder and B. van de Velde-Schlick, 2 vols., Dordrecht: Reidel (1978/1979).

E. Schröder (1966), Vorlesungen über die Algebra der Logik (1890), vol. I, New York: Chelsea Publishing Company.

G. Tonelli (1987), Da Leibniz a Kant. Saggi sul pensiero del Settecento, a cura di C. Cesa, Napoli: Prismi.

J. Trabant (1990), Traditionen Humboldts, Frankfurt am Main: Suhrkamp.

A. Trendelenburg (1857), Ueber Leibnizens Entwurf einer allgemeinen Charakteristik, in: Trendelenburg (1867), vol. III, pp. 1-47.

A. Trendelenburg (1867), Historische Beiträge zur Philosophie, vol. III, Berlin: Bethge.

H. von Helmholtz (1903), Vorträge und Reden, 2 vols., Braunschweig: Vieweg und Sohn.

H. von Helmholtz (1962), Treatise on Physiological Optics (1856–1867), 2 vols., edited by J.P.C. Southall, New York: Dover Publications.

H. von Helmholtz (1977), Epistemological Writings. The Paul Hertz/Moritz Schlick Centenary Edition of 1921 with Notes and Commentary by the Editors, newly transl. by M.F. Low, edited with an Introduction and Bibliography by R.S. Cohen and Y. Elkana, Dordrecht/Boston: Reidel Publishing Company.

A.N. Whitehead (1958), Symbolism. Its Meaning and Effect (1928), London: Cambridge University Press.

L. Wittgenstein (1961), Notebooks (1914–1916), edited by G. von Wright and G.E. Ascombe, Oxford: Blackwell.

L. Wittgenstein (1974), *Tractatus* logico-philosophicus (1921), transl. by D.F. Pears and B.F. McGuinness, London: Routledge and Kegan Paul.

2. On the Use and Character of Symbols in Modern Physical Theories

Ion-Olimpiu Stamatescu

"Der Herr, dessen das Orakel zu Delphi ist, offenbart nicht und verbirgt nicht, sondern kündet in Zeichen."[1]

1 Some Remarks About Physics

1.1 On the Physicist's Part in Epistemological Discussions

We shall venture here into a discussion of some aspects of modern physics from the point of view of the use of symbols. Every physicist would agree that what physical knowledge is dealing with, from observations up to theories, are signs – namely such signs, or symbols, whose particularity is to relate concepts which can be bound in mathematical structures to "rules" for describing phenomena (for instance, Hermann von Helmholtz: "Perceptions are for our consciousness signs whose meaning is to be learned by our reason"[2]). However, the subsequent question of whether these symbols are somehow "told" to us, are discovered or are free intersubjective constructs leads to a multiplicity of positions.[3] This suggests a rather complex discussion, and we refer the reader to the other chapters in this book to find a detailed treatment of various aspects of this problem. For more general philosophical considerations concerning symbols in science see especially Chaps. 1, 3 and 4. Here we shall take the simple–minded physicist point of view.

Physicists are sometimes sloppy in advancing philosophical interpretations of their findings. Their contribution to the discussion is, however, significant in two respects: firstly, they know that what they primarily are responsible for are their findings, even if this makes them inaccurate in their epistemological interpretations; and secondly, they are directly involved in the intricate

---

[1] "The lord, whose oracle is in Delphi, does not reveal and does not conceal but announces through signs." This quotation from Heraclit is used by Hermann Weyl to introduce a discussion on the philosophy of natural science (see Weyl 1928). Here and throughout the translations are ad hoc.

[2] "Die Sinnesempfindungen sind für unser Bewußtsein Zeichen, deren Bedeutung verstehen zu lernen, unserem Verstande überlassen ist" (Helmholtz 1896). In the following we shall speak of physical symbols and of physical concepts without carefully differentiating between them. For a careful discussion of the concept of symbol see Chap. 1.

[3] To quote only two, classical ones, see the contrast between Helmholtz and Hertz, briefly mentioned at the beginning of Sect. 2.1.

process of the development of physical ideas. We shall here therefore not attempt philosophical accuracy but proceed from the point of view of physics and try to describe some features of its symbolic structure. (Also there are, of course, many opinions; we believe, however, that this discussion is to a large extent representative of the way physicists tend to think about these matters).

Comment 1: It is not even clear whether physics supports a clear cut philosophical discussion in the sense of it being completely analyzable in terms of any one philosophical system. Typically in its development physics would leave each time behind it the world view which it itself has helped to build up – for instance, the Newtonian space–time as a a priori objectification frame, or classical mechanics as a guarantee for determinism and continuity. While the dialog between physics and philosophy is necessary and fruitful, physics would not do its job if its first concern were to agree with the latter. To quote Einstein:

> "The scientist ... cannot afford to carry his striving for epistemological systematics [too] far. He accepts gratefully the epistemological conceptual analysis; but the external conditions, which are set for him by the facts of experience, do not permit him to let himself be too much restricted in the construction of his conceptual world by the adherence to an epistemological system."[4]

In this context philosophy of physics should not be understood as a Procustean procedure, but as an attempt to identify philosophical questions in the conceptual development of physics.

1.2   On the Requirements upon Physics

It is an essential feature of physics to be concerned not only with constructing theories as *self-consistent systems of concepts* (Heisenberg), but with constructing them under the stress of successfully describing the phenomena, that is, identifying redundancies and reducing the complexity of the phenomena to a few fundamental concepts and relations, such as to be able to control the former (make predictions, etc.). This means that physics is constantly confronted with an "alien element", and its trial to account for this confrontation is to build increasingly powerful theories realizing a "transmutation" of empirical information into theoretical structures based on mathematical schemes. This continuous process of transmutation determines the character of the symbolic networks of physics. Since before (or without) theory the alien is not tractable and after (or within) theory there is no alien left, this suggests that a discussion about the role and character of symbols must be made while keeping in mind that experiment, models and theories are bound together in an intricate process of *development* of physical knowledge. This development is constrained between the demands of empirical fitness and of theoretical–mathematical self-consistency.

---

[4] A. Einstein, "Reply to criticisms", quoted in Cushing (1998), p. 357.

Comment 2: The proper consideration of the confrontation with the alien element (see, for example, Quine, 1969) and of the transmutation process above does not depend on metaphysical or other points of view; it is essential, however, in order to make contact with physics. As with any human endeavour, physics has its history and social embedding. It surely is not easy to see how, at any given moment, physical understanding stays connected to the general level of human performance and understanding, and at the same time is set up to build context-independent knowledge. Fashionable relativistic trends, for instance, try sometimes to overtax the social aspect. Such views, however, instead of "integrating physics into society", tend in fact to lose contact with it, since, because of a lack of understanding of the confrontation pointed at above and its associated conceptual development, they confuse social realization of the research process with the traits of its results. As interesting in itself as the historic–social aspect doubtlessly is, it is, however, the subtlety of the interaction between the research process and the development of physical knowledge which provides the most exciting challenge for the theory (and history) of science considerations.

The features of the above transmutation process have been discussed many times – for an example see Chap. II in Cassirer (1935). For a comprehensive review of various stands hereto see, e. g., Cushing (1998). We shall here only comment on the two aspects mentioned above, the way the new is approached and how it is incorporated in the theory.

"The theory alone decides what is observable and what not."[5] Empirical information always needs a theoretical frame to be spelled out; this is usually given by the existing theories. If the facts, however, do not fit in this frame one starts extending models which are already floating around or developing new ones, setting up phenomenological rules to describe the observations without reducing them to known theories or just using analogies, making hypotheses, investigating new mathematical structures, etc., with the aim of achieving a new theoretical frame in which the new observations can be associated with *observables* well defined as concepts in this new frame. This is specific to all major developments. A typical example is the history of the *neutrino*, a weakly interacting, neutral and therefore very evasive particle. Its history begins with the observation (by Chadwick, in 1919) of irregularities in the energy record of the β-decay of the neutron, where a proton and an electron were found as decay products. These observations were not only in quantitative but even in qualitative disagreement with the theoretical expectations. The very fundamental laws of energy and momentum conservation imply for a "one-into-two" decay fixed energies of the decay products; however, these energies appeared continuously. The ensuing discussion did not even spare the energy conservation law – although the hypothesis of its violation was never considered a serious candidate. Further results concerning, for instance, angular momentum balance were also obtained; then in 1930 Wolfgang Pauli suggested the existence of a very unusual, practically massless, spin $\frac{1}{2}$ and only weakly interacting, neutral particle, to solve the puzzle (its participation in the decay would not be directly detected, but would affects the other participants the right way to agree with the observations). The ensuing development of the theoretical scheme could incorpo-

---

[5] "Erst die Theorie entscheidet darüber, was man beobachten kann." Albert Einstein, quoted in "einem Gespräch mit Werner Heisenberg 1926" (see Heisenberg, 1969).

rate this hypothesis so well that its existence was widely accepted, although only 1953 (now specially designed) experimental search could produce direct empirical evidence for the neutrino. In a sense one can say that the current theory of electro-weak interactions started with the discovery of the neutrino half a century before. But this theory could only be established when a solid theoretical–mathematical scheme could be constructed in the 1970's.

One should notice here the "gradient" typical for these situations: we start with a theoretical framework in which we can ask some empirical question, and obtain an unexpected answer which *transcends* this framework and prompts hypotheses beyond (possibly, contradicting) it. The theoretical framework then changes to accommodate this answer. Significant progress is achieved when the old theoretical framework was "tight" enough not to permit simple redesigning and when we succeed in working out a new tight theoretical framework.

Now, from the point of view of a given theory, interpreted experimental facts and theoretical predictions are conceptually on the same footing. The "eight-fold-way" (quark) model of Gell-Mann, Ne'eman and Zweig, for instance, was proposed in 1962 on the basis of observations concerning symmetries and hierarchies of known particles. It did its job in accounting for (*postdicting*) these observations, but the mathematical structure accounting for these symmetries implied further relations and it was found out that it also *predicted* the existence of a new particle (the so called Ω-baryon), which was then found experimentally 2 years later. This was considered as a major confirmation for the model, since predictions are less prone to bias than postdictions – but from the point of view of the model itself the Ω-baryon was no puzzle. Of course, the theory may not be complete or may introduce its own "unknowns" – irreducible, empirically fixed parameters or concepts which from the point of view of this particular theory appear *ad hoc*. But that part of the observational connections which is reproducible from the theory contains no alien elements, besides those acknowledged by the theory itself. This looks like "we could have known it all from the beginning", which, of course, is a fallacy, since we had to guess the theory (here, specifically, the symmetry group) in the first place, using the previous observations, and, even if the theory is correct, it also only represents one step (see discussion in Sect. 3).

1.3   On the Dynamics of Physics Development

Physics research is a dynamical process, which is very natural considering its hypothetical proceeding. This feature, however, is not immediately apparent from the textbook descriptions which present the particular theories within their logical structure and the logical relations between these structures. For an understanding it is necessary to leave out the – sometimes tortuous – path leading to the establishment of a theory, since the student needs to learn the full conceptual structure in its achieved form – in order to be able to build upon it. However, if one wants to speak about the development of theories, one should not forget that this part exists. This does not mean that we always need to follow the detail of the tortuous path which realizes (historically) the relation between theories: we only should be aware of the fact that this realization existed and rely on it, when necessary, for finding out conceptual connections.

In a different way theory of science sometimes also seems to "freeze" physics in non-communicating, opposing sectors. Again, this is a justified view if one wants to catch the differences between the old and new concepts and make evident the change in our physical understanding. However, if one asks about the motivation and direction of the change, again one cannot disregard the processual character of the development. Notions such as "revolution" (Kuhn, 1962) offer a frame for recognizing the steps in the evolution of physical knowledge; they may, however, misleadingly suggest a high degree of arbitrariness and contingency if the conceptual dynamics is neglected.

Physics research is on the one hand an intellectual adventure, since it generates conceptual structures. On the other hand physics lives from successfully meeting the confrontation with the alien element mentioned above. The latter is primordial, as can be seen, e.g., from the fact that physics would transcend any closed, self-consistent system of concepts if this confrontation would require it. Of course empirical knowledge at each stage is influenced by the existent theoretical background – see Comment 2. The fact is, however, that our ideas seem to be steadily forced into some direction, and this independently of what we want or not to think. Consider, for instance, the developments in the electrodynamics a century ago. The ether concept had for a while a strong hold on the theoretical development, and it appeared intuitive to assume an elastic medium supporting the electromagnetic waves. Nevertheless the failure of models based on the ether hypothesis to provide a systematic frame to account for the various experimental results made the ether hypothesis more and more cumbersome and finally led to a change in the concepts. On the other hand, this change would have been difficult (and therefore it would be difficult to understand how it happened) if there would not have been at the same time other ideas already under discussion (see Chap. 6, see also Cushing, 1998). This is a subtle process. Steven Weinberg, for instance, says in his Nobel prize lecture 1979 (Weinberg, 1980), concerning the establishment of the electro-weak theory:

> "At times, our efforts are illuminated by a brilliant experiment, such as the 1973 discovery of neutral current neutrino reactions. But even in the dark times between experimental breakthroughs, there always continues a steady evolution of theoretical ideas, leading almost imperceptibly to changes in previous beliefs."

Theoretical and experimental developments are two intermingled processes, interacting steadily but also retaining a certain individual status. Surely enough, experiments prompt theoretical proposals and help to shape theories, and theoretical progress influences the design and interpretation of experimental tests. But there will always appear hypotheses not prompted by an experiment and experimental results preceding a theoretical enquiry.

Comment 3: The development of the *standard model of fundamental phenomena* with its particle "families" is an example of steady interaction between experiment and theory. But, for instance, the introduction of electromagnetic potentials in electrodynamics has not been prompted by any experiment. Conversely, the history of modern elementary particle physics, for instance, is full of discoveries, sometimes incidental, waiting for decades to be accounted for theoretically (see Comment 2).

The establishment of a theory is often a compelling event by which a number of outstanding questions are settled and a sense of conceptual completion sets in. Such was the advent of the relativity theories and of quantum mechanics. A more recent case is the rise of quantum chromodynamics (QCD). Although already developed to a successful theory for electromagnetic phenomena by as early as 1949 (QED), quantum field theory has for a long time not succeeded to take over in elementary particle physics. This was due both to the lack of systematic understanding of the mathematical difficulties in defining the theory and of their treatment (such that, e. g., Dirac would still see the renormalization procedure, which deals with these mathematical aspects, as only a provisional recipe and "the remarkable agreement between its results and experiment ... as a fluke" – Dirac, 1963) and to the lack of a dynamical concept which would explain the observed phenomena. In the preface to a well–known text book, two renowned physicists wrote in 1964 "The unsatisfactory status of present-day elementary particle theory does not allow one [the] luxury [to advocate] any single view to the exclusion of the others." And therefore they proposed to start from the Feynman diagrams as "rules of calculations" which summarize the quantum field theory but should be developed "independently of the field theory formalism which in time may come to be viewed more as a superstructure than as a foundation" (Bjorken and Drell, 1965). In fact already at that time the colour degree of freedom for the quarks proposed some years before was being introduced; 1966 saw the first formulation of their dynamics, 1970 the proofs of renormalizability and 1974 the final formulation of the strong interaction theory, the quantum field theory of chromodynamics. Although for a number of desired effects we still only have indications but no water-tight proofs, many important tests and predictions have been obtained. Due to its conceptual consistency and its empirical success, QCD has established itself in the last two decades as a fundamental theory of strong interactions: It rules the phenomena at hadronic and larger scales, and any "deeper" theory to be set at subhadronic scale is expected to reproduce QCD at hadronic and larger scales.

1.4   Questions of the Dynamics of the Symbolic Structures

Viewing the dynamics of physics as determined by the two "forces", the inherent development of the concepts and the empirical stress induced by the confrontation with the alien element, we should like to ask

– What kind of development can be observed for the symbolic structure?
– How is empirical information involved in this development?

The evolution of the symbolic structures of physics is shaped by the demands of empirical fitness and of theoretical self-consistency. Since the latter depends on mathematical schemes which need precize structures in order to be closed, as already noted by Duhem theories typically have a rather high degree of rigidity and cannot be easily "readapted" to meet contradicting observations. We may therefore ask about the interplay of continuous and discountinuous developments of concepts and theories. Thereby it is essential to establish the effectiveness of the empirical "confrontation moment" in promoting new knowledge, in particular in dealing with incommensurable competing theories (as discussed, e. g., in Chap. 6) and to understand to what extent and in what sense the conceptual development provides bridges over the gaps (as discussed in various chapters of this book). Finally, since these developments take place in a complex intersubjective process, we also should recall the role of other factors intervening here and defining its historicity.

We shall proceed from the basis of the physics of the 20th century. This discussion has, of course, no ambition to offer history or philosophy of science considerations, such as are found in Cushing (1998), Papineau (1996), Aronson et al. (1995) – to mention only a few, recent books – and in the studies provided in other chapters of this book. Instead, its intention is to present a physicist's perspective on the use of symbols in physics. In the next section will shall make some remarks concerning physical symbols. Then we shall proceed to a brief review of modern physical theories (Sect. 3). In Sect. 4 we shall discuss the question of the character of physical knowledge and of scientific progress.

2 Some Remarks Concerning Physical Symbols

2.1 Symbols and Things

Let us take as a starting point the well-known description of the intervening of symbols in physics given by Hertz:

> "We construct internal appearances or symbols of external objects, and we make them such that what results by thought-necessity from such symbols will always be a symbol of that, what follows by nature-necessity from the symbolized objects [...] – The symbols we speak of are our representations for things; they have with the things the one essential concordance which consists of satisfying the above requirement, but it is not necessary for their scope to have any other concordance with the things."[6]

---

[6] "Wir machen uns innere Scheinbilder oder Symbole der äußeren Gegenstände, und zwar machen wir sie von solcher Art, daß die denknotwendigen Folgen der Bilder stets wieder Bilder seien von den naturnotwendigen Folgen der abgebildeten Gegenstände [...] – Die Bilder, von welchen wir reden, sind unsere Vorstellungen von den Dingen; sie haben mit den Dingen die eine wesentliche Übereinstimmung, welche in der Erfüllung der genannten Forderung liegt, aber es ist für ihren Zweck nicht nötig, daß sie irgend eine weitere Übereinstimmung mit den Dingen hätten." (see Hertz, 1894).

This view may be contrasted with that of Helmholtz, who does not mention a constructive aspect in establishing of symbols.[7]

Hertz makes two aspects equally important: The *stringency* of the symbolic construction (coming from the reference to "things" and from the mapping obtained in this way of the "nature-necessity" by the "thought-necessity") and the *freedom* in choosing the symbols. We shall use these remarks as a basis for a brief discussion of: the valences of physical symbols, the status of the symbols and of their associations to "things" and the establishment of physical concepts.

Comment 4: It may be interesting at this point to ask what kind of "things" do we encounter in physics, to which our symbols are supposed to refer in some sense. Firstly, we have things which, at least in some theoretical context and in some approximation, are recognized as "material" objects: bodies, particles, fields, strings, and so on. Secondly, we encounter concepts referring to objects with no apparent "materiality", like force, curved space–time or wave functions. Finally our concepts refer sometimes to even more abstract kinds of things, which can be identified as peculiar patterns of behaviour of more concrete ones: waves, flow types, fixed points, attractors, solitons, etc. We may deny such "non-material objects" the character of things. The delimitations, however, are not always very clear: so, for instance, "force" in the modern theories results from the local interaction with a field which carries energy (e.g., the photon) and is therefore, in this sense, "material"; soliton-type of excitations of some fields are assumed to manifest themselves as particles, while such fundamental fields as quarks are themselves not observable directly. We should also note that in many cases "things" which are well identified mathematically may have a counterpart in nature, but hidden in the complexity of phenomena – this is typically the case for the "reference" of such concepts like strange attractors and "catastrophes", but it also holds, for instance, for quarks. Of course, one may try to use such criteria as mechanical properties (e.g. the capacity to carry energy and momentum), complexity, stability, etc., to distinguish between, say, things of predominantly "matter" character and of predominantly "rule" character. Generally, and observing the developments in physics in the last 150 years or so, we are tempted to associate with the concept of "thing" *some* capacity to support properties and to retain its identity in different contexts – as a particle, but also a field, a soliton, or curved space would do – but not as just being a mere

---

[7] In direct continuation of the text quoted in footnote 2 we read: "Wenn wir jene Symbole richtig zu lesen gelernt haben, so sind wir imstande, mit ihrer Hilfe unsere Handlungen so einzurichten, dass dieselben den gewünschten Erfolg haben, d.h. dass die erwarteten neuen Sinnesempfindungen eintreten." ("When we have learned to read these symbols correctly, we are able with their help to shape our acts in such a way that the latter have the desired success, i.e., the expected new perceptions occur.") Even if read metaphorically this statement seems to accentuate a passive attitude, in contradistinction with the view of Hertz. Notice, however, that the latter speaks of "objects" and "things" as given, not constructed. It seems therefore that Hertz only pushes the passive level of "something being found" to a lower stage. For a detailed discussion of Hertz's conception of physical symbols, see Chaps. 1, 5 and 8. See also Dosch (1997).

universal relation – such as, for instance, the rules for Lorentz transformations, a conservation law or the Schrödinger equation. See also Sect. 2.4. (One should not try to immediately identify in the above the philosophical concepts of substance and function. Such an interesting but complex discussion is not intended here.)

2.2   The Valences of the Physical Symbols

The double stringency noticed above reflects a double valence (and correspondingly, function) of the physical symbols: "horizontal", within the *formal, mathematical structure*, which represents the relations between concepts (e.g., definitions, kinematical constraints, dynamical equations, etc.); and "vertical", within the *interpretational structure* which essentially collects the rules relating the mathematical symbols to phenomena (notice that we do not mean here the epistemological interpretation). Normally, the immediate relation of a symbol is to another symbol (for a discussion of the semantic network of physical symbols see Chap. 10). Nevertheless any interpretational scheme ends with an experimental or observational setup and a protocol through which the "alien element" shows up in what we call empirical information.

The capability to participate in the "vertical" chain of relations processing the "empirical" information (connecting to the external world) and the capability to participate in the "horizontal" network of the theory (which is based only on its internal, mathematical logics), represent separate valences of the physical symbols, and they form together the basis of the stringency of the symbolic structures.

Comment 5: So, for instance, Maxwell equations relate the symbols for electric and magnetic fields, their variations over space and time, charge densities, etc. The time variation of the magnetic field, for example, equates a combination of variations of the electric field along spatial directions. These are horizontal correlations in our terminology. The magnetic field itself can come from a magnet, and a change in the field at some point may be produced by moving the magnet. Likewise, an electric field will move charges (as implied by the horizontal correlations) and produce a current in a wire. Currents and magnets enter directly an experimental setup – or an observation upon nature: they can be defined by actions on other bodies, acting as probes. Surely enough every stage of the vertical chain (magnets, needles, currents) is again defined with the help of symbols which participate in the network of the theory (but usually can be defined more generally, the lower the stage in the chain at which they act: we do not need the full Maxwell equations to introduce magnets, e.g.). It is incontestable, however, that at the end of the chain something is acting which is prior to the symbols, and which will force us to change the symbols and their correlations if they do not match the observations. For this reason we consider the two types of correlations (horizontal and vertical) as different.

2.3   On the Status of the Symbols

We do not doubt that "there is" something of which we speak as "the solar system", and this something did not change when we changed our concept of it (but what exactly we called "the solar system" changed, both in representation and in reference: what orbits, which planets, etc.). Also, the existence of life on earth depends (among others) on something coming from the sun we call "light"; we described the way this thing comes and acts the way it does increasingly well in mythology, in classical physics and in quantum physics. What can we now say about the status of the symbols of a theory which appears to have been selected to account for certain phenomena (planets, gravitational force, electromagnetic fields, photons)? Notice that this is not a question about the "existence" of mathematical entities. The physical symbols are not reducible to mathematical signs: they select and use the latter but need to comply with further requirements.

One suggestion may be that when we ask about the "existence" of things, concerning the symbols the correct notion to ask about would be that of "necessity". This is a stronger notion than "adequacy", and it appears justified by the exclusive character of selecting among competing theories, on the one hand, and by the analytic necessities within the conceptual network of a theory, on the other hand. While the latter is essentially a logical (mathematical) question, the former is more ambiguous. In fact we witness long periods of coexistence of competing models or even theories and sometimes the choice is not indisputable, or it appears justified only later due to new insights, or it is overtaken at a further level of understanding. (As examples of these three cases we may think of quantum *versus* Bohm's mechanics, minimal interaction in quantum electrodynamics, and the particle and wave character of light between Newton and Einstein.) Nevertheless, *in the long run* these "fluctuations" do not seem destabilizing; on the contrary, they appear to help ensure open-mindedness and facilitate an evolution in the course of which ambiguities tend to be resolved (see also Sect. 1.3). It thus appears that, once we begin to deal with very many phenomena and ask for comprehensive conceptual structures, the *stringency* implied by Hertz's "requirement" compels the *freedom* on such narrow tracks that we become tempted to speak of "necessity" in introducing physical symbols. We shall see later, however, that this necessity cannot be understood in a strong sense and it needs qualification.

Comment 6: Although this point does not directly concern our main line of argument, it may be interesting to briefly discuss it here. Firstly, of course, we have genuine "covariances" inside theories, such as Schrödinger, Heisenberg or path integral formulations in quantum mechanics, or the alternative formulations of electrodynamics with help of retarded/advanced potentials instead of electromagnetic fields. The different schemes introduce different symbols, and these call on different "objects". Nevertheless we mostly observe three situations: either heuristic arguments, which usually will be either justified or contradicted by later insights, select one of the representations (this is the case in classical electrodynamics, where the field representation was chosen), or the representations appear also conceptually equivalent (the various pictures in quantum mechanics, which can be transformed into each other in the frame of the theory), or they appear adequate according to the physical conditions (such as particle or field representation in quantum field theory). In all cases, however, the "covariances" appear themselves meaningful as such, in that they help find equivalences and introduce relevance criteria.

Secondly, we have the case of the so-called empirical under-determination of theories. A typical example might be that of Bohm's mechanics as an alternative to quantum mechanics. Here, however, the empirical equivalence is proven only for rather simple situations, Bohm's mechanics, which can be understood as a *non-local hidden variable theory* soon becomes rather cumbersome, and it does not appear prone to relativistic generalization (quantum field theory). Therefore it is not clear whether one can speak here of real empirical under-determination, especially if we consider the development potential. In other cases, such as Lorentz, Einstein and Abraham electrodynamics, for instance, the under-determination is only temporary. A modern example is that of the *local hidden variable theories* or of the *spontaneous collapse models* for quantum mechanics, where the situation is or can be empirically settled. This problem must be therefore analyzed carefully in each case.

Notice that the choice between theories is not always "minimalistic". A good criterion, for instance, is usually that the theory should not introduce *ad hoc* parameters or unjustified supplementary quantities. However, this is a delicate aspect. We may remember, for example, the role of the electromagnetic potentials in electrodynamics, which generate the (observable) fields but also introduce redundant (gauge, unobservable) degrees of freedom: While in the classical theory their use could only be justified for reasons such as "elegance", it turned out that a consistent development to a quantum theory needed the potentials.

2.4   On the Association of Symbols to "Things"

Concerning this let us first ask which could be the "objects" ("die Gegenstände") of which Hertz speaks? (It is irrelevant at this point whether we think of them as "condensations" of observations or as something behind, and showing up through, the observations. In any case we shall mean by "objects" the things such as they are recognized through our concepts.) We can say, for example, that the best way to account for a certain class of phenomena is to speak of an electromagnetic field – allowing at the same time that the precise way of speaking of it may change to account for new observations. Now, if the "objects" are simply "the observed phenomena" ("Beobachtungen"), then we are free to introduce independent symbols, say, for light and for electromagnetic waves. And this is adequate under certain conditions – e.g., if we are only concerned with linear optics or with radio waves. However, at a deeper level of understanding this would be wrong – and we may remember the beautiful discussion of Maxwell concerning the identity of light with electromagnetic waves[8]. This is no longer a question of choosing, because some effects will only be predicted correctly under the assumption that light *is* a high–frequency electromagnetic wave. This means that the "objects" are

---

[8]   J.C. Maxwell, *Scientific Papers*, vol. I, Cambridge 1890, p. 526 ff, reproduced in Samburski (1975), p. 560 ff.

neither directly observed phenomena, or just non-obliging names for contingent correlations between them, nor arbitrary theoretical constructs set by convention to represent the phenomena, but they are built in the development of physical understanding.

Thus this association has two directions: on the one hand it suggests appropriate concepts, and on the other hand it helps to identify objects. Notice that this happens in a process involving "gradients" related to the empirical stress (see Comment 2) and to conceptual drifts (neglecting this aspect one may be tempted to see here only the closed hermeneutic circles which stabilize the various concepts, and wonder about their ambiguities – see, for example, M. Hesse, "Models, Metaphors and Truth", in Radman, 1995, p. 353). Equating light and electromagnetic waves is a genuine theoretical step – it is a non-trivial and fruitful hypothesis, not just a renaming.

On the other hand, these gradients also indicate that the above association cannot be a fixed relation.

Comment 7: Of course one can call $Electron^{ED}$ the thing electrodynamics speaks of and $Electron^{QED}$ the thing quantum electrodynamics speaks of, and they are (more or less) *fixed* in the respective symbolic networks; but then they can only be said to point approximately to one thing in the reality, since the "objects" they *define* do not coincide. The interesting point is, however, that these $Electron^{TM}$ appear as marks on a path which seems "to be there" at least for a while, and which we shall simply call $\mathcal{E}lectron$. The "track" $\mathcal{E}lectron$ represents a directed path, in the sense in which one theory overrides the other. Clearly the symbol $\mathcal{E}lectron$ is of a different and much more metaphoric character than $Electron^{ED}$ or $Electron^{QED}$ (since it correlates to both of the latter in an unprecise way, and in fact suggests that there may appear further concepts to which it may correlate some day). Nevertheless it shows a necessity which in some sense transcends that of $Electron^{ED}$ or $Electron^{QED}$ because it seems to hold beyond the binding in a particular symbolic network: a conceptual path such as $\mathcal{E}lectron$ seems more directly related to the "alien element" which we need to deal with when we proceed in our symbolic construction, since this has influenced us in setting the different marks. However, without the marks we cannot recognize that there is a path (but see also Sect. 3.5).

Of course, the above descriptions seem to call for the notion of metaphor. However, the kind of metaphors appearing here need very many explanations if one wants to avoid equivocations. So, for instance, one can consider the statement "light is an electromagnetic wave" itself as a metaphor, since it relates two apparently disparate phenomena. However, the phenomena are disparate only from the point of view of their observation; once we learned enough about them this is no longer a metaphor but an equivalence statement. Other metaphors (such as the analogy between sound and light) may trigger the wrong heuristics (ether) and may be turned down finally. Also seeing models as metaphors, for instance, needs many clarifications. (For further discussion see, e.g., E. Monteschi, "What is wrong with talking of metaphors in science?", in Radman, 1995; Miller, 1996.) Since the purpose of this article is to provide material for discussion, and not interpretations, we shall not attempt such an analysis.

2.5   The Constructive Aspect of the Necessity in the Association Symbols-to-Things

We cannot (without further metaphysical assumptions) base the necessities apparent in the "transmutation" process of empirical information in conceptual structures on some more fundamental concept. But this does not mean renouncing the requirement for necessity; on the contrary, it seems to be an important stabilizing factor in a process based on "proposing and testing hypotheses", since it forces us to be more restrictive than mere "adequacy" would require. Also the observed development of science seems to support some concept of necessity in choosing theories and producing hierarchies. We should only be careful that the relations implied by this "necessity" cannot be simple and immovable. One possible view of this necessity is that it should be understood constructively: partly based on the foregoing network of symbols, partly on the new empirical information and partly on the further analytical constraints imposed by the new, ensuing theory. In this "marriage" the empirical moment has the upper hand: if required, the theoretical networks of connections will be changed. This happened to the ether, to the Galilean concept of simultaneity, to the classical concept of particles, and as we already notice from these three examples, there are various degrees of change. In some cases some concepts which have proved wrong were disposed of: they were unfruitful in the sense of blocking further development (ether). In other cases they were replaced by new ones, of which they may represent an "approximation": they are neutral (Galilean simultaneity). Finally they may be changed by "enrichment"; they are fruitful in helping us to develop new theories (particles). Necessities and associations appear here to be graded and qualified. It seems, therefore, that it may be meaningful to speak of necessity, but only in some "weak" sense.

Comment 8: Of course, besides new empirical information there can be also the observation of "logical" contradictions in the old network which forced the change. The "necessity" revealed in this step is partly reducible to the one described before: usually the weak points have been spared in the old theory because they were not relevant phenomenologically and there is refined or new empirical information to make them relevant. Nevertheless this points to the fact that we are often forced to deal with models and theoretical constructions which do not represent closed systems of concepts in a strong sense, and where therefore also the formal necessities are incomplete or at least not fully ascertained.

2.6   On the Establishment of Physical Concepts

It is apparent that there are at least two ways to speak of the establishment of physical concepts and symbols: the first one concerns their logical (interpretational and mathematical) binding, the other one is their concrete history. At the former level a number of more or less systematic procedures play an important role. They concern the concrete steps undertaken in establishing both the vertical chains and the horizontal structures mentioned above – idealizations, approximations, accessing new mathematical structures, analytic and numerical calculations, instantiations in models, data processing, etc. (see also Chap. 8). At the second level we may expect an interplay between "logical" conclusiveness (where we also include induction and hypothesis) and randomness and various other factors which introduce contingencies. This interplay is well described in works concerned with the history of physics, for an example see Kuhn (1978). Among others, here is a moment where also philosophical convictions and world views can be active. A quoted example is the positivist reserve of Kaufmann, who in 1911 had measured properties of the electron at the same time and more precisely than Thomson, but would not relate his results to the concept of a particle and therefore would not develop his enquiries. An even more prominent example in the same spirit is the conventionalist attitude of Poincaré, who had the mathematical scheme of special relativity before Einstein, however would not propose it as a basic theory but only as a computational scheme. On the other hand, these examples strengthen in fact the stringency of the theoretical development, since they show its high degree of independence – *in the long run* – on factors external to physics. The concept of electron, for instance, established itself in the form the contemporary level of understanding would best permit it (classical physics, quantum mechanics), and would later change again in the wake of new insights (quantum field theory), without regard to its epistemological status. Similarly, the installation of Einstein's special relativity was not hindered by epistemological differences and both the mathematical and the interpretational structure put forward by it do not depend on this controversy.

Finally, in between these levels is the strange level, where such criteria as: beauty, generative power, simplicity, etc., play a role. Let us consider, as one among a multitude of examples, the minimal interaction in quantum electrodynamics, the assumption that the interaction between charges and electromagnetic fields are represented by the simplest mathematical construct connecting their symbols, namely their local product. The simplicity of this hypothesis was one heuristic reason for it to be chosen among some possible ones. On the other hand, it has turned out afterwards that this form of the interaction is necessary to ensure the renormalizability of the theory, which itself proved a very general and powerful criterion for the mathematical and interpretational consistency of quantum field theories. It seems therefore that even here there is little place for chance, especially if one remembers that sooner or later all conjectures and hypotheses have to face the requirement of mathematical and empirical consistency. Nevertheless one must acknowledge that at least for a while such heuristic criteria are active for themselves, i.e. without further justification. See also Miller (1996).

With regard to the *intersubjective* process (which emerges in, but is not reducible to the social context) concerning the symbolic structures of physics, we can observe: Search for agreement about a dominant theoretical frame (on the basis of both conceptual and empirical arguments), together with questioning of its validity (on the same basis and with possibly decisive effects), and the presence of a significant "subdominant" area of models and ideas which may never but sometimes do come to realization and become relevant. A more detailed sketch of this process can be found, e.g., in Th. Kuhn's (1962) description (which, although partial, has the merit of stressing the positive role of the interaction between old and new); another view is presented, for instance, in Beller (1999). An optimistic view of this process (e.g., in Peirce's perspective of convergence of the intersubjective process "toward a unique limit", or in the more reserved perspective of Cassirer's "invariants of knowledge") could find here a constructive balance between inertial and dynamical traits. At least the experience until now does not contradict this view.

3 Modern Fundamental Physical Theories and Their Relations

This section recalls some features of the actual theoretical architecture of physics. We apologize for its length. We shall start with some remarks on the structure of the research landscape.

3.1 The Structure of Physics Research

In discussing the symbolic structures of physics, one is tempted to consider the relational and interpretational networks of fully fledged theories. However the actual procedure in physics is much more complicated and involves a nearly endless zoo of models, conjectures and isolated hypotheses, empirical laws and uninterpreted experimental results which participate in constructing the meaning of the physical symbols. Since the mature meanings are essentially taken over into the established theories and live in the symbolic structures of the latter, looking at the theories to identify physical symbols is not incorrect. Nevertheless, if we want to characterize the dynamics of the conceptual evolutions, it is useful to understand its realization conditions and be able to follow it in actual realizations.

As an example for uninterpreted experimental results we may remember the discovery of the neutrino mentioned in Comment 2, examples for empirical laws are the Balmer, Lyman and Paschen spectral series, which were first accounted for theoretically by Bohr's atom model and can be said to have contributed in prompting quantum mechanics. Conjectures and isolated hypotheses are frequently precursors of models or are made in the frame of the latter (so, for instance, the hypothesis of quantized orbits in Bohr's model). In the following we shall concentrate on the role of models.

Modeling in physics always takes place in a certain theoretical background (sometimes very general, or primitive) and either refers in some sense to an existing theory (or set of theories) or presupposes the forthcoming achievement of such a theory. An example of the first kind is the Rutherford atom model (classical revolution of electrons in the Coulomb field of the nucleus; this model made evident the inconsistency of classical physics, because for such a motion classical electrodynamics predicts loss of energy of the electrons by continuous radiation and hence their fall on the nucleus). Other examples of the first kind are the statistical mechanics models (which are simple examples of statistical mechanics systems and which permit the study of phase transitions, critical behaviour, etc.), the cosmological models (which are based on Einstein's equations of general relativity), etc. Examples of the second kind are provided by the Bohr model for an atom (classical orbits, but quantized angular momentum) or the Landau model of superconductivity (effective equation for an order parameter). Notice that a model does not necessarily rely on some classical, intuitive picture as a basic input (as in the case of the mentioned atom models): it can just as well consist in an equation or other abstract formulae (such as the Landau model of superconductivity). A model also should not be understood as an approximation to reality. A model is in itself a simplified "artificial reality", established in some background of theoretical understanding (which can be very primitive) and open to our enquiry (including numerical simulations). Models are typically less stringent than theories and therefore more flexible than them (but also less "robust"). They are directly involved in the interpretation of theories and in the dynamics of theory development. They may also establish by themselves a framework for research. From the point of view of our discussion it is important to note the role of models in generating symbols. Sometimes such symbols are precursors of the mature symbols taken up in the ensuing theories (Bohr's atom, quark model), and at other times they are more *ad hoc* and model-dependent (mean field, bag model, etc.). In both cases, however, the multiplicity and variability of the concepts promoted by the models should be understood in relation to the modeling framework, either as trials in the attempt to build up a new theoretical structure (productive or explorative function below) or as "effective" concepts developed in the framework of a theory in view of applications (interpretative or applicative function). In the first case the uncorrelated multiplicity is mostly transient, in the second one it is representative for the particularity of phenomena, without contradicting the theoretical scheme on which the models rest[9].

---

[9] This contradicts the position of Nancy Cartwright, who advocates a "patchwork" of phenomenological laws (if consequently followed, this would mean: down to each phenomenon) – see, for example, "Fundamentalism vs. The Patchwork of Laws", in Papineau, 1996. Cartwright, however, only considers the extreme po-

Comment 9: Models fulfill a number of cognitive functions, which mostly appear together in each model, but usually with different weights (we do not mean here "function" from the epistemologic point of view, but just in the sense of the role models play in the scientific process). The following functions could be identified (the author apologizes for the rather *ad hoc* designations):

Interpretative: To obtain phenomenological predictions from a theory. One hypothesizes, for instance, "effective concepts" which mediate between the abstract elements of the theory and phenomenological observables. This is especially useful in modern theories which typically do not allow easy calculations. Example: mean field in solid state physics. Here one assumes, for example, that for each atom of a solid the effects of all the others (attractions, repulsions) average to a "mean field" which this atom feels. This is a fictive quantity which once assumed, however, can be estimated and permits in turn observable quantities such as phase transition parameters, etc., to also be estimated. The calculation of such parameters in a mean field model represents thus an approximation; on the other hand, the assumption of a mean field, if systematically successful in providing good approximations, also indicates that the averaging by which it is introduced is itself a good approximation. The fictional "mean field" becomes "physical", i.e., even if it does not claim to represent "an element of reality" in the sense the fundamental entities in the theory to some extent do (the atoms and their elementary interactions in this example), it represents an identifiable situation. The interpretative function presupposes a theory and does not intend to go beyond this.

Applicative: An important subset of interpretative models consists of models proposed for solving classes of physical problems. The typical case is that of statistical mechanics models, set up to deal with questions such as regarding the explanation of the properties of glasses, crystalline properties, electric and magnetic properties of materials, phase transitions, chaotic behaviour, etc. The mentioned Landau model for superconductivity, for example, and the more involved "BCS" microscopic model (for which the Landau model serves as precursor) belong to this type. While these models, like the interpretative ones, do not go beyond the theory which serves them as a basis (classical statistical mechanics, quantum mechanics), the accent is here not on understanding or testing the theory but on solving important types of physical problems.

Productive: To introduce new concepts and laws which are not supported by existing theories. In the dynamics of theory development, the necessity of new insights (prompted by empirical contradictions or by internal inconsistencies of the existing theories) is seldom met by producing new theories from the beginning. An intermediary step is taken over by models which in introducing new hypotheses relax

---

sitions of a strong "fundamentalism" and of the mere incoherent multiplicity of phenomenological rules, while the view promoted here considers a dynamical hierarchical symbolic structure based on advancing and testing hypotheses. For example, the multiplicity of phenomenological ("effective") laws by which one tries to have a grip on the behaviour of *real* superconductors is in no contradiction with the fundamental quantum laws and well defined interactions which lead to superconductivity. Would there appear a relevant contradiction, the fundamental laws might be questioned at some moment – when this happened it has led to superior theories. Hence, it is, in fact, the alternative itself (fundamentalism *versus* patchwork) which seems artificial.

the condition of proving complete consistency among them and with respect to all known facts. Example: Bohr's atom model. Here the hypothesis of stable, quantized orbits is not compatible with the classical (mechanical and electrodynamical) frame in which the model is established. In the form in which it appears in the model it is also not compatible with quantum mechanics (which generally does not know orbits in the strict classical sense). Nevertheless in this model the quantization hypothesis explains not only the stability of matter but also the more detailed question of the hydrogen spectrum (the Balmer, Lyman and Paschen spectral series). The model promoted this hypothesis to a genuine new insight which led to quantum mechanics. The productive function provides elements for new theories. In a more general sense one can also speak of an *explorative* function when there is not enough control to lead the search and many directions have to be prospected. Such a situation, for instance, can be encountered in the strings and related models.

Constructive: In certain cases large classes of models build up a theoretical framework by themselves, which does not need to be condensed into a new fundamental theory as such. This holds, for instance, for the whole field of the *physics of complex systems*: material science models, neural networks, deterministic chaos, theoretical biology, self-organized criticality, econophysics, etc. Here the underlying theory is simply mechanics or statistical mechanics, differential equations and discrete maps, or stochastic processes. The complex system modeling is neither concerned with "interpreting" these theories (statistical mechanics, etc.), nor with developing a superior or unifying fundamental theory. In some sense their function is applicative, since they use these known theories to understand classes of phenomena and resort to the conceptual frameworks of the latter. However, in another sense they also have a productive function, since they develop a self-consistent conceptual scheme (state space, dynamical flows, attractors, fractals, chaotic behaviour, bifurcations, etc.) and aim at finding regularities in nature, which are not related to fundamental elementary aspects but to complexity and its mathematical description. They build up in this way a research field with its own theoretical framework and attempt to construct a unifying symbolic structure across the diversity of the application fields. (Of course, we could regard these models as biological or economical models, etc., but then we fail to perceive their generality. In fact, due to these universal traits they provide new concepts and methods for the fields in which they are applied, and enrich the symbolic structures of the latter, while profiting themselves from this interaction – a typical example are neural network models.)

Explanative: To help to bind together theoretical elements such as to provide a logically easy to follow scheme of arguments. Typical examples are the *Gedankenexperimente* – such as the "double slit experiment". This function appeals to various mechanisms, such as visualization, reduction, segmentation and recomposition, which generally act in the shaping of new "intuitions" (see also Sect. 4).

3.2   Established Physical Theories

The physical theories which represent our up to date physical knowledge are the classical theories of mechanics, thermodynamics, electrodynamics, statistical mechanics, special relativity, general relativity and their quantum theoretical developments: quantum mechanics and quantum field theory (while quantum gravitation is still being searched for). They can also be viewed in larger study frames, which call upon many of them simultaneously: the structure of matter, space and time, complex systems, fundamental interactions, etc. They cover more or less without gaps the whole known physical universe, from the largest to the smallest *now observable* scales. This does not mean that they explain everything: not only it is not meaningful to require them to explain life, etc., but also many physical phenomena remain yet unexplained, and we do not know whether the envisageable development can deal with them. What is meant here is that within their validity domains they do not seem to be contradicted, and that these domains are not disjunct. They may refer to a certain class of phenomena (e.g., electromagnetic, gravitational) or they may be defined as approximations (to other theories) for phenomena fulfilling some conditions (non-relativistic theories for velocities much lower than c, non-quantum theories for "decohering" situations) – but within these limits they are understood as generally valid.

Since the usual reduction goes from macroscopic to microscopic, a theory typically is valid for all length scales larger (or energy scales smaller) than those where it was established, while from the point of view of the theory set for smaller scales the former may appear as "effective" or approximate. The *standard model of elementary particles*, for instance, appears valid at all scales between subnuclear (hadronic) and cosmological ones (up to uncertainties such as those concerning the so-called "dark matter", etc., which cannot yet be assessed). On the other hand, it is expected to be only an "effective" theory from the point of view of the more fundamental, *grand unified theories*, which we hope to develop to describe the phenomena at scales much smaller than the hadronic ones.

It may seem that the reversed view also holds: thermodynamical quantities and laws, for instance, emerge when we deal with very many degrees of freedom, some peculiar regularities, as flow patterns and so on only appear at large scales. But this does not mean that mechanics, e.g., is not valid there (it is just consequent application of microsopic laws which leads to such peculiar behaviour – "deterministic chaos", for instance). It indicates, however, the explanative limitations of the reduction of such phenomena at these larger scales to microscopic laws, and the necessity of developing concepts directly dealing with complexity.

Comment 10: It is sometimes stated that quantum mechanics is a theory for *microscopic* phenomena. This is somewhat misleading. In fact, quantum mechanics is a non-relativistic theory of particles and therefore limited from the start to phenomena at scales *larger* (or energies *smaller*) than those at which pair creation may set in (roughly, 10^{-13} cm, or $1\,\mathrm{MeV}$), and this also means that it will not apply to phenomena ruled, e.g., by *strong interactions*. The confusion arises because quantum mechanics is typically *relevant* for microscopic phenomena (above the scale of strong interactions), while most macroscopic phenomena can be satisfactorily described by classical laws. However, quantum mechanics claims validity and is shown to apply also to the macroscopic mechanical phenomena, both in

treating special cases where quantum effects are evident at a macroscopic scale –
superconductivity, for instance, – and in explaining the classical observations typical in most other macroscopic situations (but also in some microscopic ones), as
we know from the study of *decoherence* effects (see next section). In this latter case
the laws of behaviour predicted by quantum mechanics are *for all practical purposes* just the classical ones.[10] As a formal theoretical frame, quantum theory, like
thermodynamics, appears to provide a universal type of description.

3.3 Relations Between Theories

Among the above theories there are a number of interesting relations. We
shall briefly mention:

(a) the microscopic grounding of thermodynamics on (statistical) mechanics;
(b) the developments of mechanics to relativistic mechanics and to quantum
    mechanics;
(c) the unification of gravity and space–time in the frame of the general
    relativity theory;
(d) the development of classical field theory (electrodynamics) and quantum
    mechanics to quantum field theory.

(a) Thermodynamics is a "phenomenological" theory concerning the description of phenomena with the help of *extensive* quantities, such as energy
and entropy, and of *intensive* parameters, such as temperature. An extensive
quantity will add when we put together more systems, an intensive one will
average in a certain way until equilibrium between systems is achieved. A
thermodynamic analysis will identify an "energy", an "entropy" and a "temperature", among other quantities, and relate them with the help of three
principles and of the theorems derivable from them. Statistical mechanics
pertains to the macroscopic description of systems containing many microscopic, fluctuating degrees of freedom (variables). Thereby "global" quantities
can be defined which "collect" contributions from the microscopic degrees of
freedom and we can again identify an "energy", an "entropy" and a "temperature". Historically, thermodynamics has been developed for phenomena
concerning heat, but the frame is much more general. Likewise, statistical
mechanics came first as a theory of gases, but it also provides a very general
frame in which any "collaboration" of many microscopic degrees of freedom

---

[10] Here and elsewhere we shall mean by *valid for all practical purposes* that no
feasible experiment or observation can establish a deviation. In particular, in
the above case, to observe interference effects contradicting the classical behaviour of macroscopic bodies, say, one would need to use the whole universe as
apparatus. The difference from an "impossibility of principle" remains then of
at most cosmological relevance. We may recall that similar arguments have been
exchanged a century before concerning irreversibility in statistical mechanics –
Poincaré cycles, etc.

can be treated in view of the overall ("macroscopic") behaviour. These two description frames are related, in that the thermodynamic quantities can be defined with the help of the statistical mechanics ones. Although the principles of thermodynamics cannot be strictly *derived* from statistical mechanics (that is, without further assumptions about treating microscopic information – coarse graining, *Stoßzahlansatz* – and about initial conditions), the latter explains them in terms of more fundamental concepts and makes them therefore more transparent. If, for instance, we cannot "deduce" the irreversibility, we can at least see which are the conditions which may produce it, instead of just postulating it. In this sense we have here an example of reduction in which the (partially) reduced theory is conceptually enriched by the reduction and simultaneously strengthens the significance of the symbols of the more elementary theory, while the latter, although proving itself especially in the reduction of the phenomenological theory, goes beyond the frame of that one.

(b) A common sentence is that it is impossible to predict in which position a glass slowly pushed over the edge of a table will reach the ground. This, however, is significant for the problem of stability of solutions and the so–called "sensitive dependence on initial conditions" – and not for the correctness of *classical mechanics*. Notice that for a cat in the same situation one can predict that it will fall on its legs, which means that it can (albeit not trivially) control its spinning movement without probably any help from some cat-daemon. On the other hand we know that classical mechanics is incorrect for describing fast moving bodies, which makes it approximative. What one should note, however, is that here *approximative* does not mean an arbitrary choice or point of view, but the identification of a non-equivocal physical situation: slow movement is defined in terms of a *classical mechanics* quantity (velocity) with the help of a limiting phenomenon (light) – without appeal to the whole structure of mechanics (and even less so to the *special relativity theory*). One can identify in a consistent way the set of phenomena for which classical mechanics holds; to quantify the "goodness" of the approximation, however, we need relativistic mechanics.

Special Relativity Theory (SRT) is based on two principle: (i) the equivalence of physical laws in all inertial reference frames (in uniform motion relative to each other), and (ii) the observed finiteness of the velocity of light and its independence on the reference frame (these hold for any massless field). SRT leads to an essential change in the concepts of space and time. Although the lack of Galilean invariance in *electrodynamics* – Maxwell's equations are not invariant under a change of reference frame which assumes an absolute time – had already raised many questions and had promoted a vivid development of ideas, it needed a stringent construction, provided by SRT, to accept that *simultaneity* is a relative – and in this sense unphysical – concept. We have here therefore an example of a theory overriding another one by proving that the bases of the other are false, but at the same time giving the latter a well–defined status as an approximation with a wide domain of applicability. Notice, however, that the concepts of the "old" theory are never really reproduced in the limit, but replaced *for all practical purposes* by "simplified" (approximative) forms of the new, richer concepts. See also Sect. 3.4 and Chap. 6.

The relation between classical and quantum mechanics is more subtle. First of all we have the question of the role of classical concepts in establishing the conceptual scheme of quantum mechanics. However, this is mainly an epistemological problem, while we are for the moment interested in physical–theoretical questions. We shall therefore stay in the frame of quantum mechanics and ask where we find classical mechanics here. The answer is that we cannot find it as such, i.e., we cannot obtain the original, classical symbolic structure in some limit. However, we can determine in the frame of quantum mechanics which the phenomena are for which classical mechanics offers a correct description *for all practical purposes*. More precisely, we can determine for given physical conditions whether our observations of the system under considerations will or will not permit quantum–mechanical interference effects to be identified – in the last case we speak of "decoherence". Quite generally, classical equations of motion hold for certain quantum–mechanical expectation values, therefore under decoherence conditions our observations will attest to classical behaviour, since neither interference effects nor deviations from classical motion will be observable. We stress that this is a quantum–mechanical effect. It is due to the *quantum mechanical–correlations* with other systems with which our system unavoidably interacts. If those other systems leave the region of observation – for instance the light reflecting on our system – quantum-mechanical coherence quickly becomes delocalized. Then all observations *of the system considered* will show no effects of this typical quantum-mechanical coherence, since for this to show up we must also observe the whole environment; the latter, however, is no longer available. An experiment measuring, for instance, the position of the system will find it classically localized within a small region and will show no quantum–mechanical interferences. It turns out, as it should, that this is typically the case for macroscopic objects, since they have multiple interactions with the environment, but it also explains a number of microscopic phenomena (e.g., *chiral molecules*) for which one can show that exactly the kind of conditions hold for which the above-described "decoherence effect" acts.[11]

Comment 11: For an illustration of decoherence effects: air at room temperature and pressure would localize a larger molecule of size 10^{-7} cm (10 Å) to within its own size in about 10^{-14} s. Subjecting this molecule under these conditions to a double–slit experiment of resolution worse than the above would show it to behave classically. Decreasing the pressure down to laboratory vacuum would increase the

---

[11] For an introduction see Zurek (1991), for a systematic study Giulini et al. (1996).

localization time to about $\frac{1}{10}$ s, and this would permit the double–slit experiment, which before would have attested to a classical behaviour of the molecule, to bring now into evidence interference effects and thus show the quantum behaviour of the same molecule. Collisions in a dense medium will make even such typical quantum objects as protons and electrons behave classically, leaving, for instance, "classical" traces in Wilson chambers. For a dust particle of size 10^{-3} cm in perfect vacuum (empty space), even the extremely weak cosmic background radiation would localize it to within its radius in less than a second, and smooth, thermal radiation at room temperature would localize it in 10^{-13} s. The larger objects of daily experience are bound to behave classically under practically all circumstances.

Of course, as already mentioned, decoherence does not replace the measurement postulate (in fact, it rests on it). Here is a deeper, nearly century–old epistemological problem, related to the question of "making sense" of the symbols of the theory, and represented in the theoretical scheme by the measurement postulate (see also Chap. 9). We first should notice that what is meant here is the question of "making *classical* sense" [12]. The difficulty of this program has brought many people to desperation and triggered the positivist attitude especially represented by Born. However, the quantum–mechanical probabilities (or whatever) have to pertain in some way to the observed systems (or whatever); otherwise we do not know of what we are speaking. The so–called probabilistic interpretation is indeed very simply correct, since it just does not put forward such questions but solves them by fiat: we do not want to know what electrons have to do with state vectors, but we know that measurements on electrons are predictable from state vectors. (The situation is the same if instead of a "system" – here, electrons – one speaks of a "reproducible experimental setup".) It seems that we still have to live with the situation as it is: a direct assignment (the "descriptive" stance: the situation of an electron is completely described by the state vector) seems contradictory in view of the measurement postulate; the simple solution of "know nothing but here is the result" (the theory is not assumed to pertain to the events of the actual experiment, but only to the statistics of their results) [13] does not take full advantage of the potentiality of the symbolic structure of the theory (state vectors, superposition principle, decoherence mechanisms, etc.). The situation remains therefore a challenge both for physics and for the philosophy of physics.

(c) General Relativity Theory (GRT) is a beautiful construction which unifies two apparently unrelated theoretical schemes (gravity and space–time) into one self-consistent theory. The concept of space–time entering GRT is that of a continuous structure, a differentiable manifold (up to various types of singularities); it is, however, more general than the space–time of Newton's theory or of Special Relativity Theory (SRT) in that it can show an (es-

---

[12] And therefore might only be "forced on us by our language, a language that evolved in a world governed very nearly by classical physics.", Weinberg (1993), p. 85.

[13] This is in some sense a "black box" stance: The actual experiment is made up of individual events – the sparks on the screen, say – which can be taken at arbitrary time intervals one after the other. The probabilistic interpretation tells us what the cumulative result is but does not allow us to ask how this result is to be obtained from the individual events.

sentially smooth) internal deformation called curvature. Curved spaces have long been known in mathematics (Gauss, Bolyai, Lobatchewsky, Riemann), and physicists such as Helmholtz considered it to be an empirical question whether physical space supports a non-Euclidean geometry. But it took SRT for us to realize that one should not consider space and time separately but as the compound *space–time*.

Gravitation was identified by Newton as being an attractive force between material bodies, which acts at a distance, and whose strength is proportional to the masses of the bodies and inversely proportional to the (square of the) distance between them. Using the equations of motion and this force could explain the laws of falling bodies as well as the revolutions of planets. An essential ingredient in this derivation is the assumed identity of the "inertial mass" entering Newton's equations as the proportionality factor between force and acceleration, with the "heavy mass" determining the strength of the gravitational force – the so-called Galilean "equivalence principle" (resulting from the statement that all bodies fall with the same speed).

GRT is based on two principles: (i) general covariance – the laws of physics do not depend on the description (coordinates); and (ii) Galilean equivalence, which means that gravitation fields can be replaced *locally* by accelerated reference systems (the continuity properties of space–time are essential for this). Therefore, GRT generalizes the relativity principle of SRT from inertial systems to any systems. There is no force at a distance in GRT, but only local influences and propagation from point to point. Instead of an absolute space–time which cannot be acted upon *and* a gravitation force at a distance which governs the motion of masses, we obtain a *unified concept* of a space–time whose internal properties (curvature) rule the motion of bodies on geodesics (shortest paths), and in this way reproduce a gravitational interaction among them.

(d) Classical electrodynamics is a relativistic theory of charges and of space–time distributions, the electric and magnetic fields: Besides electromagnetic fields and charges there are no elementary forces or other primary objects in electrodynamics – everything is derived from the former. Quantum mechanics is a non–relativistic theory of microscopic bodies interacting by forces which are given from outside the theory. In particular charged particles will interact by (classical) electromagnetic forces (neglecting relativistic effects). It turns out that both the quantization of electrodynamics and the special relativistic development of the quantum mechanics of charged particles in interaction with electromagnetic fields lead to the same theory, the quantum field theory of electrodynamics. This is a much more powerful theory than the other two and introduces new concepts and relations. Its fundamental concept, the quantum field, is achieved by developing the previous concepts of (quantum–mechanical) particles and (classical, relativistic) fields. We see here therefore that the common application of the two principles, the quantum and the (special) relativistic ones, leads to the "unification" of two theories in a superior one which has a richer conceptual structure than the simple "addition" of the earlier conceptual schemes.

Comment 12: All physical theories introduce measurable entities ("quantities" expressible as real numbers by comparison with some units) and relations among them. If in the frame of established theories there are relations between various physical quantities which hold under all circumstances, these quantities become "commensurable" by just these relations. This happens, for example, between time and space (through the Lorentz transformations of special relativity), between energy and temperature (in thermodynamics and statistical mechanics) or between energy and frequency (in quantum mechanics)[14]. The translation parameters introduced by these relations – the speed of light, c, the Boltzmann constant, k_B and Planck's constant, $\hbar$ – have conventional values which only reflect the fact that we have chosen to measure our quantities using beforehand given units: seconds, meters, degrees, joules, etc. Conversely, we can choose to set these three *fundamental constants* all to 1, by which we only acknowledge that these quantities are commensurable and therefore there is only one fundamental *dimension*, which we may choose, say, to be the length dimension. A second means then just $\simeq 3 \times 10^8$ m, a joule $\simeq 3.3 \times 10^{25}$ m^{-1}, and a speed of 108 km/h, say, is given simply as $\simeq 10^{-7}$ (meaning this fraction of the speed of light). In the framework of the current knowledge it seems conceivable that one unifying theory could exist which would combine all current theories and explain these and all remaining relations (which are just ratios between various quantities: the masses of various particles, the temperatures of phase transitions, the electromagnetic properties, etc.). It is less clear whether a theory can exist in which the last fundamental constant (say, Planck's length $l_P \simeq 10^{-35}$ m) becomes derivable. We may dream of a final theory, but this might be better understood as regulative idea; see also Weinberg, 1993.

Planck's length illustrates the problem of quantum gravity. As far as quantum physics is valid, for a particle of mass M quantum effects become important at distances $R_Q \simeq 1/M$ (its "Compton wavelength", which is a measure of the localizability of the particle). If general relativity is valid there, the Schwarzschild horizon is $R_S \simeq GM$ (where G is the gravitation constant). Hence, trying to improve localization (decrease R_Q by increasing M) we may fall within the increasing black-hole horizon R_S. This defines Planck's length, $l_P = 1/M_P = \sqrt{G}$, as the best localization we can achieve, because an attempt to improve from the point of view of one theory is contradicted by the other theory. This calls therefore for a unified treatment.

For an overview of the current "leading-edge" research in physics see Chap. 10.

---

[14] This does not mean that these quantities become identified: energy and temperature, for instance, have very different conceptual statuses and the meaning of the relation between them *belongs* to the theoretical framework of statistical mechanics.

3.4 Evolution of Symbols

In the four examples of the previous section we encountered three situations (reduction, overriding and unification) and correspondingly various kinds of symbol evolutions.

Consider the reduction of thermodynamics to statistical mechanics. Energy, entropy and temperature, for instance, are defined in thermodynamics at two levels: once in the abstract scheme, axiomatically, and once in relation to observations, phenomenologically. The abstract scheme can be applied to different types of phenomena and catches in itself a certain kind of relationships which can be generally observed in nature, such as conservation (of energy) or exchange (of work and heat). Thus these symbols show two aspects: on the one hand, they serve to identify these special relationships; on the other hand, they can represent observables and instantiate these abstract relations. Statistical mechanics proceeds from microscopic, mechanical quantities (e.g., kinetic energies of air molecules) and from statistical procedures (counting the number of microscopic states of a large number of microscopic degrees of freedom, e.g., the moving molecules) to define global, macroscopic observables (average energy, etc.). When it was shown that energy, entropy and temperature as defined in statistical mechanics follow the laws of thermodynamics, the thermodynamic symbols acquired a new strength ("necessity"), since the question of relating them to phenomena turned into a systematic procedure; the symbols of statistical mechanics were similarly "upgraded" by being shown to give rise to relations of a very general kind; and the physical understanding gained insight into the relations between microscopic and macroscopic features.

Consider now the overriding of non–relativistic by the relativistic mechanics, and let us discuss the difference, say, between Einsteinian and Newtonian mass (see Chap. 6). Newtonian mechanics is in fact not discarded from physics as being an incorrect theory; on the contrary it is redefined (and thus "secured") in the frame of relativistic mechanics as a well–defined *approximation*, as noticed above. Under the conditions which permit this approximation, the Einsteinian "rest mass" is not only a kinematic parameter but becomes directly dynamically relevant, as proportionality factor between force and acceleration. In fact, in relativity theory the relativistic mass itself is in some sense a "derived" quantity; the variables fixing the physical situation are the rest mass (a kinematic parameter given as an "invariant" under a symmetry, the Poincaré group; a procedure which may be regarded as an effective way of defining "kinds" of observables) and the velocity with respect to observer; everything else is obtained from these two. It is now interesting to compare the start and end points of the circuit *classical Newtonian mechanics → relativistic mechanics → $v/c = 0$ limit of relativistic mechanics*. At the end of this circuit the Newtonian inertial mass, recognized as dynamically relevant property of a particle, is "recovered" from the rest mass, which now appears at its place in the equation of motion. The recovered concept, however, is not *identical* with the original (Newtonian) one, but is enriched by the latent significations coming from relativistic physics (the reference to Lorentz invariance, for instance, or even quantitatively, as prescriptions for correction terms). It seems that we see here how the conceptual changes accompanying the development from classical mechanics to relativistic mechanics (α) introduce new, irreducible elements and (β) do not mean irreplaceably discarding the old concepts but subjecting them to shifts.

In the other overriding case (non-relativistic classical mechanics $\to$ non-relativistic quantum mechanics) the situation is more complicated. Here the approximation is not defined in terms of one kinematic parameter (the velocity) but in terms of complex physical situations which involve the dynamics in an essential way (decoherence is a dynamical effect). A fundamental symbol such as "particle" when followed in a similar circuit as above shows a measurable difference to the original, classical symbol, which cannot always be made arbitrarily small. Moreover there are symbols in quantum theory which have no classical target – the wave function, for instance (remember that even in decohering situations there is still a wave function for the system together with the environment; it is only that this entanglement has no consequences for observations upon the system). This means that the recovered (decohered) theory is not the classical theory. Nevertheless, besides the agreement in predictions, those symbols of the theory which can be put into relation operationally also fulfill the same functional relations and agree to a certain extent concerning their attributes. So, for instance, a decohered quantum particle, which can be observed as a classical particle, also obeys the same equations of motion as the latter, and shows within some limits the same properties in a wide domain of cases. The classical symbols therefore do not become useless; it is only that, while the decohered particle can be followed back continuously to a fully quantum phenomenon (even operationally, by changing the experimental conditions – see Comment 11), this is logically impossible starting from the classical concept[15]. We thus observe a variation on the (α) − (β) theme above.

Finally, in the unification case we observe how the symbols of the earlier theories not only evolve but also melt together. For instance, in GRT, both the Newtonian absolute space and time, which do not react to what happens *in them*, and the uncanny "action at a distance" (which Leibniz, because of its non-causal character, considered a "return into the kingdom of darkness"[16]) vanish from the theoretical structure. They are replaced by the unifying concept of a space–time whose internal properties (curvature) reproduce a gravitational interaction. The former concepts can be recovered as approximations after a number of steps (small masses, flat space–time,

---

[15] In fact, historically, the development of quantum mechanics proceeded by extending the classical concepts – the "correspondence principle", for instance –, but this appeared only as recipe with the "new" residing in the "extension rules".

[16] "Le royaume des ténèbres" – see Cassirer (1935), p. 208.

small velocities, approximate Galilean relativity), but the GRT space–time concept is much richer and is not exhausted in these limiting concepts.

Likewise, the concept of a particle (which is still fundamental in quantum mechanics, even if it is not the perfectly localizable, identifiable classical object) and the concept of a field become in quantum field theory just two ways of manifestation of the same fundamental object, the quantum field. Here we no longer have conservation of particles (as in quantum mechanics), and their properties (mass, charge, etc) are determined through the symmetry and interaction structure of the theory; the fields lose the smooth character of continuous distributions (as in classical electromagnetism), and in their high–energy interaction a discontinuous, particle character is apparent. Again we can in some sense regain the earlier concepts in some limits, but the rich structure of the quantum field is no longer visible there.

This aspect of "overriding within unification" appears rather specific to the modern theoretical development (see also next section).

Comment 13: Notice that the conceptual unification reached in the quantum field is not that concerning the *particle–wave complementarity* in quantum mechanics. In the latter case we start from the symbol of the classical particle and construct the quantum particle by "superposing" a "wave function" onto the classical particle describing its measurable properties (position, momentum, spin, etc.). In a decohered situation the "classical" aspect shows up (we see the particle), and the quantum character is relegated to the entanglement with the environment and remains unobservable. There is no real, classical wave associated with the wave function in some limit (if we do not resort to hidden variables).

The quantum field can be constructed in two ways, which we shall exemplify here for quantum electrodynamics:

(a) We can start from a classical field, e.g. the electromagnetic one, but instead of considering its values at all space-time points to be exactly correlated by the laws of the theory (the Maxwell equations), we assume that the field can fluctuate and we introduce a wave function over its values at each point (in much the same way as we give a wave function over the values of the positions of a quantum-mechanical particle, instead of fixing these positions by the Newton equations). It then turns out that the so-quantized field exhibits at short distances (high energies) a granular, particle character (the photons).

(b) We start from non-relativistic, quantum-mechanical particles (electrons), but assume relativistic energy–momentum relations for them. It then turns out that anti-particles (positrons) must also exist and that electrons and positrons can annihilate; hence the particle number is not conserved. We can introduce creation and annihilation operators to represent these quantum particles at a given momentum, spin, etc., and combine them into the "frequency" (mode) representation of a quantum field. The latter then satisfies relativistic equations of motion, which are the quantum equivalent of certain wave equations.

What a "classical limit" can obtain from the quantum field depends on the type of field. So, for instance, zero-mass fields can support long–distance correlations and appear in the classical limit as smooth, classical waves (electromagnetic waves, for instance). Massive fields, on the other hand, can be observed in decohered situations as classical particles.

Notice also that the equations of motion satisfied by the quantum fields, and which describe their dynamics, can have non-linear terms as long as these do not violate other conditions (locality, renormalizability, etc.). In contrast to this, the quantum–mechanical wave function of a particle satisfies the strictly linear Schrödinger equation (excepting for some "collapse models", which violate quantum mechanics).

3.5 On the Symbols of Modern Physical Theories

The symbols acting in present–day physics appear to have many faces when we want to give them meanings – both in the sense of describing the relations among them and in trying to develop an intuition for them. The latter has to do with the attempt to achieve "pictures" allowing simultaneous grasping of multiple relations and features – as an image would do. It seems characteristic of present–day science that on the one hand this is a necessary step in meaning giving; on the other hand it only can be achieved by contrasting many "pictures", where each of which taken alone is not only partial but in fact false (see, e.g., Stamatescu 1995, see also Chap. 7). Since there are many levels of abstraction which our intuitions have to catch up with in trying to get "familiar" with these symbols, this situation should not be too surprising. See also Sect. 4.3.

However, also the "function" of the symbols, as represented by the relations inside the symbolic networks, both "horizontal" and "vertical", apparently becomes multiply faceted, with similar symbols fulfilling different relations in hierarchically organized structures. So, for instance, the fundamental concepts of space and time have changed, not only in their internal structure (from Galilean to relativistic physics) but even in their role. While in classical (Galilean and special relativistic) physics they are description frames, they become dynamic degrees of freedom in gravitation (general relativity) theory, or become intermingled with the dynamics in their short–distance (continuum) structure in special relativistic quantum field theory (see Stamatescu, 1994, 1998). Similar things happen with all other important concepts, such as fields, particles, interactions, charges, mass and energy. Typically, different symbols appear no longer to be independent but to represent only different realizations of some more fundamental, unifying concept – see, e.g., the quantum field discussed in Comment 13. From a unifying theory of gravitation and of the other fundamental interactions, one expects in fact that even particles and fields, on the one hand, and space–time, on the other hand, will also appear as only different facets of some more fundamental concept (e.g., strings) which will also fix uniquely the various symmetries and thus the interaction. Charge, mass and so on emerge from symmetries (unitary symmetries of the gauge degrees of freedom, Lorentz symmetry of space–time transformations, etc.).

Following the evolution of the symbolic structures of physics, we seem to witness two intermixed kinds of developments of the symbols and their function. On the one hand, we have the evolution of principles which generate the relations structuring the symbolic networks, on the other hand, we have the evolution of the "objects" defined through the symbols. Examples of the first type concern the relativity and covariance principles, the role of symmetries, quantization (from minimal action to path integral), etc. Examples of the second kind concern particles, fields, space–time.

These developments seem to produce hierarchies in which some theories and concepts appear to accept a "redefinition" (*for all practical purposes*) as "effective", approximate realizations of some more fundamental ones established at a smaller scale. This proceeds, for instance, by averaging microscopic degrees of freedom (statistical mechanics $\rightarrow$ thermodynamics), by decoherence of phases (quantum mechanics $\rightarrow$ classical mechanics) or by symmetry breaking as a result of varying some parameter (grand unified theories $\rightarrow$ standard model).

Asking in this context questions such as that of "necessity" and that of the "association to things" (or "definition of objects"), we notice that these notions still seem to hold, but to depend on some gross conditions, such as the scale of the phenomena considered. So not only do, say, $Electron^{ED}$ and $Electron^{QED}$ show such conditional necessities and associations, but even the fuzzier concept $\mathcal{E}lectron$, which seemed to stay behind them, may not find further realizations at smaller scales. Nonetheless, even if conditional, these necessities and associations are neither arbitrary nor disconnected across the borders of their validity regions. Thus, for instance, high–energy experiments use decohering situations (e.g., a bubble chamber or more involved detectors) to identify the quantum particles (which are created in processes described in quantum field theory) by their classical behaviour (leaving curved traces in the chamber with some magnetic field applied): a continuous chain from relativistic quantum field theory over quantum mechanics to classical mechanics. The effective degrees of freedom emerging via phase transitions from higher symmetric "objects" represent both the new conditions and the objects they originated from. Therefore this "multiply faceted" character does not appear to weaken the coherence of the symbolic scheme.

We should also notice that modern science appears to emphasize a certain trend in the formation of symbolic structures, namely the establishment of symbols proceeding from mathematical models. This goes beyond the definition of symbols with the help of the formal structure of the theory (such as wave function or $Electron^{QED}$) or the prediction of phenomena on the basis of the mathematical implications from the theory – such as the prediction of particles, of phase transitions or of black holes. We witness, for instance in the theory of complex systems, symbolic networks developed starting from certain mathematical schemes involved in the modeling of simple interaction structures, such as competing forces, dissipation, collective effects, etc.

The symbols introduced in this context (cycles, attractors, catastrophes, etc.) pertain primarily to general properties of complex system modeling and less to well-defined classes of phenomena. They acknowledge their mathematical genealogy by representing relationships and patterns of behaviour and by being easily transportable (we speak of attractors in pattern recognition as well as in population evolution). The descriptions they provide of the phenomena of the real world are model-dependent, therefore the necessity character associated with them is primarily mathematical, while their empirical necessity is more difficult to assess. Some of these concepts pertain only to the modeling problem – such is the case for convergence properties of particular expansions, for validity boundaries identified in the frame of particular models, for types of critical behaviour, etc. Other concepts may have referents in the "reality", which, however, may be masked behind the complexity of the phenomena. For example, attractors in artificial neural networks are very "natural", since most dynamics show this kind of behaviour; it is much more difficult to identify them in natural networks (e.g. brains). This does not mean that observational instances are missing – solitary waves, fractals, etc. have been introduced on the basis of observations – but that the identification of their role in a complex phenomenon is sometimes difficult.

4   On Physical Knowledge and Scientific Progress

4.1   The "Pragmatic" Attitude of Physicists

We have seen some aspects of the way physics proceeds in forming its symbolic structures, and some of the characteristics of the latter. A question which we did not touch is: How can we characterize the kind of knowledge we arrive at in this way?

We apologize again for not attempting philosophical accuracy and for not trying to take the standpoint of one or other philosophical system. In fact, from the point of view of physics we tend to see epistemological arguments as less contradictory than they may appear once they are bound in different philosophical systems. As already noticed, while physics requires stringency concerning both mathematical consistency and experimental conditions, it is much less purist concerning philosophical aspects. In the latter case it rather proceeds by neglecting contradictions than by choosing one point of view.

This pragmatic attitude starts at the realist/idealist alternative, which we shall use here for a paradigmatic illustration of this point: "In natural science the opposing world views of Realism and Idealism designate non-contradictory methodological principles.... We construct [in natural science] an objective world in which simultaneously two principles must hold: A 'realistic' principle [which, following Helmholtz, could be described as follows] – 'a difference in the perceptions reaching us is always due to a difference of the real conditions.'... [further], an 'idealistic' principle – 'the objective picture of the world should allow no differences which could not show up in perception; an existence which by principle is closed to perception is not accepted.' "[17]

The discussion about realist, idealist, positivist, etc., traits of physical understanding is a very vivid and interesting one. For classical physics one could have said, with Helmholtz, that a "realist *hypothesis*" (that the world of material things exists independently of our conception) "[is] the simplest we can put forward, tested and confirmed in extraordinary wide domains of applications, sharply defined in all details and therefore extraordinary useful and fruitful as a basis for handling." [18]

Further insights and observations forced us, however, to develop quantum theories, and in these theories the "material world" Helmholtz is speaking about is no longer a well defined concept throughout (it is at best an "effective concept" in the sense of decoherence). It is still true that we must and apparently can count on the understandability of the world, whether we state that as regularity with Helmholtz or, more precisely, as a "general principle of causality" ("allgemeiner Kausalsatz") with Ernst Cassirer[19], or

---

[17] "Innerhalb der Naturwissenschaft bezeichnen die weltanschaulichen Gegensätze von Realismus und Idealismus einander nicht widersprechende methodische Prinzipien.... Wir konstruieren in ihr eine objektive Welt, in der zugleich zwei Prinzipien gelten müssen: Ein 'realistisches' Prinzip, [das man mit Helmholtz so darstellen kann]: 'Eine Verschiedenheit der sich uns aufdrängenden Wahrnehmungen ist stets in einer Verschiedenheit der reellen Bedingungen fundiert.'... [Ferner] ein 'idealistisches' Prinzip: 'das objektive Weltbild darf keine Verschiedenheiten zulassen, die nicht in Verschiedenheiten der Wahrnehmung sich kundgeben können; ein prinzipiell der Wahrnehmung unzugängliches Sein wird nicht zugestanden.' " (Weyl, 1976, p. 84.

[18] [Die realistische Hypothese] "sieht als unabhängig von von unserem Vorstellen bstehend an, ... die materielle Welt außer uns. ... [Sie ist] die einfachste [Hypothese], die wir bilden können, geprüft und bestätigt in außerordentlich weiten Kreisen der Anwendung, scharf definiert in allen Einzelbestimmungen und deshalb außerordentlich brauchbar und fruchtbar als Grundlage für das Handeln." "But," he continues, "we cannot acknowledge it to be more than a hypothesis" – "für mehr als eine ausgezeichnet brauchbare und präzise Hypothese können wir die realistische Meinung nicht anerkennen." (Helmholtz, 1878, p. 273).

[19] Which he considers to represent, conceptually, "a jump in empty space" ("ein Sprung im Nichts"): "... [Der allgemeinen Kausalsatz] kann nur als 'transzendentale Aussage' verstanden werden, die sich nicht sowohl auf Gegenstände als vielmehr auf unsere Erkenntnis von Gegenständen überhaupt bezieht. ... das Suchen nach immer allgemeineren Gesetzen ist ein Grundzug, ein regulatives Prinzip unseres Denkens. Eben dieses regulative Prinzip, und nichts anderes, ist das, was wir Kausalgesetz nennen. In diesem Sinne ist es ein a priori gegebenes, ein transzendentales Gesetz: denn ein Beweis desselben aus der Erfahrung ist nicht möglich. Aber auf der anderen Seite gilt, daß wir für seine Anwendbarkeit keine andere Bürgschaft als seinen Erfolg haben. Wir könnten in einer Welt leben, in der jedes Atom von jedem anderen verschieden wäre; in ihr wäre keinerlei Regelmäßigkeit zu finden und unsere Denktätigkeit müßte ruhen. Aber der

general postulate with Peirce[20] or simply as "greatest wonder" with Einstein. However, the question of the character of the description of these "regularities" appears rather subtle. In particular, quantum theories appear difficult to cast in clean alternatives (beyond an undemanding positivist perspective) – see, for instance, d'Espagnat (1995), Cassirer (1935), see also other chapters in this book. Instead of directly attacking this question we shall make some remarks below about *how* physics proceeds.

Physics proceeds by making hypotheses and testing them both for consistency within the conceptual (incl. mathematical) structure and for empirical fitness, which are thus the stabilizing factors in this hypothetical proceeding.[21] Two important elements here are truth and objectification. It may be interesting to illustrate for these two concepts the different perspectives of physics and of philosophy.

4.2 On Justification and Truth

A philosophical pragmatist may claim to be "suspicious about the distinction between justification and truth, for that distinction makes no difference to my decision about what to do." (Rorty, 1998, p. 19). As a physicist I am

---

Forscher rechnet nicht mit einer solchen Welt; er vertraut auf die Begreifbarkeit der Naturerscheinungen, und jeder einzelne Induktionsschluß wäre hinfällig, wenn ihm nicht dieses allgemeine Vertrauen zugrunde läge. 'Hier gilt nur der eine Rat: Vertraue und handle! - das Unzulängliche wird dann Ereignis'." ("... [the general principle of causality] can only be understood as a 'transcendental statement' which does not refer to objects, but to our knowledge of objects. ... the search for increasingly general laws is a fundamental trait, a 'regulative principle' of our thinking. It is exactly this 'regulative principle' and nothing else we call the law of causality. In this sense it is a transcendental law, given a priori: a proof of it starting from experience is not possible. But on the other hand it also holds that we have no other guarantee for its applicability than its success. We could have lived in a world in which each atom would be different from another; in this world we could find no regularities and our thinking would have to rest. But the researcher does not count on such a world; he trusts the understandability of phenomena, and each inductive conclusion would be inappropriate if it would not be based on this general faith. 'Here only holds one council: trust and perform, the inadequacy will then become event'."). (See Cassirer, 1935, p. 200; the final discussion, in particular the inner quotation, follows Helmholtz 1896, p 591 ff; see also Helmholtz, 1878, p. 278).

[20] See Peirce (1890). For Peirce the understandability of the natural process is a postulate, or, just as well, a "desperate hope", since only in as far as this holds is knowledge possible.

[21] This does not settle the question "What are hypotheses about, what is their content?" In this context one sometimes speaks of "hypothetic realism" (see, e.g., Vollmer, 1990), "structural realism" (see, e.g., J. Worrall, in Papineau, 1996), etc. Popper's three worlds theory may also be seen as an example of an attempt to answer this question – see Popper (1972).

inclined to think that surely this distinction makes no difference to me in deciding an action but may make a great deal of difference to what happens to me after performing the decided action and, if under "justification" I am allowed to understand physical theories, that:

- the anticipation of possible unexpected reactions from the environment is the motivation for research programs;[22]
- the research improves our knowledge, in that previously unexpected reactions are accommodated in new justifications;
- research programs lead *in the long run* to ordered justification structures (the previous level is either replaced or incorporated – e.g., recognized as an approximation with a well-defined domain of validity); and
- competition situations are solved *in the long run* in an "objective" way, in that the alternative with the best development capacities tends to take over (again, either by elimination or by incorporation).

In particular this is the reason why one continues to test the predictions of even established theories: the present "justification network", as solid as it may appear, can still be defectuous or can miss further connections (see, for instance, the important, present day field of quantum mechanic tests). The expectation of a possible discrepancy between our predictions (based on justifications) and the actual happenings is therefore pragmatically relevant (it makes us eager to learn), and this expectation itself hides in it an "additional norm" besides justification, since we do not expect that whatever was behind the previous discrepancy depends on our improved justification but the other way around.

Both philosophers and physicists are right within their validation criteria. But these differ since they are designed for different problems. Generally, philosophy asks for conceptual (not mathematical) stringency, while the physics view is dynamic, based on mathematical relationships and oriented toward the confrontation with the *alien element* mentioned before. Physicists tend to feel that, concerning natural science at least, respecting some norm of truth is indispensable. Even if this concept appear philosophically troublesome, it catches an aspect not covered by justification alone and is therefore pragmatically relevant.

Of course, one can redefine *justification*, the way, e.g., Peirce does it, as the infinite process of conceptual translation (in the frame of a communication process) of the increasingly successful encounters with the world. This has the advantage that we do not need then to speak of truth; however, it seems that the latter is in fact here assumed without further specification. In connection with Peirce's construction, this follows from his "postulate of intelligibility"

---

[22] We are in fact seldom in the situation of having accounted for all known facts, i.e., of having an up-to-date perfect level of justification. Typically there are a lot of not yet explained observations around and there may be a stubborn old contradiction as much as a new unexpected one to make us consider our achieved justification level – our present theories – as unsatisfactory.

(of the world) and his conception of the infinite process above, which he assumes to tend toward a unique limit. The motor of this process lies in the expectation of something *not yet justifiable* but *in principle intelligible*, and this implies in fact a notion of truth, at least in a dynamic – temporal, approximative – but robust sense.

4.3   On Intuition and A Priori

In a critical perspective objectification, meaning the procedure by which we identify "objects", involves intuition; therefore we shall try to qualify the role of intuition in physics.

For a long time the daily experiences of space and time could easily be taken as a basis providing "fundamental forms of intuition" for physics and therefore the conditions for objectification. This understanding, however, started to encounter a number of problems. It is not at all intuitive to accept that the course of time will depend on the reference frame. It is even more difficult to imagine a curved space based on Euclidean intuition (one needs at least one more Euclidean dimension – hence 5 in total – to embed a curved space into an Euclidean one, or a not very intuitive *tangential* construction). And, finally, one cannot really give any intuitive meaning to the statement that there are states of a particle which can be interpreted as it being at the same time *"here" and "there"* (and not only as *I don't know whether "here" or "there"*), where *"here"–"there"* are two positions orders of magnitude farther apart than the size of the particle; one must resort to the corresponding mathematical formalism to make sense of this by finding its measurable consequences.

Even if we reduce the fundamental forms of intuitions to ordering tendencies, with which Helmholtz might agree[23], we run into problems, since these orderings cannot be trivially achieved throughout: they become intermixed because of relativity effects and are limited by quantum effects.

For these reasons, a physicist may feel inclined to a conditional view upon intuition and following, e.g., Reichenbach (1978), renounce an intuition based on immovable fundamental forms. Avoiding the danger of "canonization of 'common-sense' " (Reichenbach, 1965, p. 73), and as a counterpart to the axiomatic procedure in developing theories, one could consider "intuition"

---

[23] "Kants Lehre von den a priori gegebenen Formen der Anschauung ist ein sehr glücklicher und klarer Ausdruck des Sachverhältnisses; aber diese Formen müssen inhaltsleer und frei genug sein, um jeden Inhalt, der überhaupt in die betreffende Form der Wahrnehmung eintreten kann, aufzunehmen." ("Kant's teaching of the a priori forms of intuition is a very clear expression of the matters: but these forms must be empty of content and free enough to take in any content which could present itself to the corresponding form of perception.") (Helmholtz, 1878, p. 299). Helmholtz supports an axiomatic setting of the geometry, as opposed to the transcendental, Kantian setting, and asks for empirical determination of the properties of the physical space.

in physics as an ever improving heuristic and interpretational *instrument*, bound itself in the dynamics of the development of physical understanding.[24]

It seems that we could see here the constructive role of the "conflicts between reason and intuition" ("Konflikte zwischen Denkkraft und Anschauen") of which Cassirer speaks quoting Goethe.[25] In this sense one could also see in the speculative "game" with mathematical models characterizing the present-day physics research, beyond the preparation of a *reservoir* of theoretical structures, the shaping of intuitions required for the next stage of physics development.

Comment 14: It does not help to argue that relativistic or quantum effects, etc. are not matter of our daily experience. The reason is twofold: Firstly, *they are* a matter of daily human experience, even if not present or not direct. Physically, biologically and (by extrapolating from the present-day possibilities) also technically it is possible, for instance, to send somebody in a spaceship accelerating steadily at $1\,g$ for some years. He may visit distant parts of the Galaxy and come back within his life time; he will then have to cope with, possibly, thousands of years having passed by here in the meantime. This is not science fiction but a possible human daily experience, such as speaking of circumnavigating the earth a few centuries before Magellan (we have assumed the correctness of special relativistic mechanics, which, however, is probably the least doubtful part of this extrapolation). Likewise, we are steadily confronted with quantum effects, even if indirectly – from superconductors, lasers and TV to the very existence of atoms (neglecting these relationships is not better justified than assuming that the milk comes from the milkman). Secondly, everyday experience is just *part* of our encounter with the world. From a critical perspective it is the theory, which was constructed observing the phenomena, which arrives to represent the means of *objectification*. Even if this has a hypothetical character, we cannot accept contradictory conditions of objectification acting simultaneously in the same object (unless, of course, we can order them in hierarchies of approximations – such as speaking, say, of approximate simultaneity for phenomena involving only "slow" movement).

4.4   On Scientific Progress

The two major achievements in physics in the last century are:

– The foundation of physics on principles of relativity and equivalence, which not only gave the theories of space and time but also led to the concept of local interaction, to a new assessment of the role of symmetries, etc.

---

[24] This is particularly evident in everyday physics research, with its continuous development and testing of partial models, both in the frame of an established theory and outside of a consistent theoretical frame, using imaging intermixed with formal arguments. See also the remarks in Sect. 3.5.

[25] In connection with the "crisis of intuition" due to quantum mechanics (see Cassirer, 1935, pp. 515, 521).

- The revealing of the basic quantum character of physical phenomena which – besides resulting in the theories of quantum mechanics and quantum fields – also allowed a new understanding of the identity and indiscernability question, a new view of the coherence of phenomena or of the question of randomness in nature, etc.

We cannot say that we have understood most of the consequences following these achievements. And we have also failed up to now to combine general relativity and quantum theory into a quantum theory of gravitation. The most important aspect of these developments, however, is the increase in unification power paralleling the raise in the abstraction level. Ernst Cassirer, for instance, considered that by accepting non-Euclidean geometries in the determination of the physical world view "the unity of the world ... not only is not destroyed, but it is in fact truly founded in a new way, since the particular laws which we have to account for in the space–time descriptions now find themselves together in the unity of a superior principle – the postulate of general relativity. The renouncement of the intuitive simplicity of the world view would thus bring with it at the same time the assurance of the larger consistency of the latter." [26] This appears to be a general trend of the symbolic structures of modern physics, although it is not clear whether we are already approaching a unified description of all physical phenomena.

One should also mention two further developments with significant influence on our understanding of the physical world, and which also prompted important developments in the research:

- In 1924 Hubble's results on cosmic distance estimations demonstrated that *nebulae* are far extragalactic objects, themselves galaxies like our own. This opening of the cosmos started the modern era in cosmology and prompted major developments in association with both General Relativity Theory and Elementary Particle Theory (the *standard cosmological model*).
- In classical physics itself the field of complex systems developed into a major theoretical framework, characterized, on the one hand, by strong relations with mathematics (information theory, complexity theory, complex maps, stochastic processes, etc.), and on the other hand by a kind of "universality" with respect to the concrete fields of application (material

---

[26] [Die Zulassung von nichteuklidischen Geometrien] "die Einheit der Welt, d.h. die Einheit unseres Erfahrungsbegriffs von einer Gesamtordnung der Phänomene nicht nur nicht zerstört, sondern sie von einer neuen Seite her erst wahrhaft begründet, indem auf diesem Wege die besonderen Naturgesetze, mit denen wir in der Raum-Zeit-Bestimmung zu rechnen haben, sich zuletzt in die Einheit eines obersten Prinzips – eben des allgemeinen Relativitätspostulats – zusammenfassen. Der Verzicht auf die anschauliche Einfachheit des Weltbildes würde also zugleich die Gewähr seiner größeren gedanklichen und systematischen Geschlossenheit in sich schließen." (Cassirer, 1921, p. 101).

science, neural and biological modeling, etc.). This is a clear demonstration of the relevance of classical physics and of its capability, without leaving its theoretical framework, to go beyond its alleged limits (simple deterministic behaviour).

Scientific progress is not meant to be simple accumulation of features but genuine development of concepts, whereby features are lost and features are won and also where conceptual "jumps" occur. Surely enough, rarely the evolution of knowledge has taken the shortest path, and sometimes we have been on the wrong path and needed to "jump". Nevertheless it does not seem as if we were all the time just arbitrarily jumping between wrong paths. On the one hand, as already noticed (see Chap. 6), the empirical basis for decision is not affected by conceptual incommensurability between competing theories. This shows that even big conceptual shifts can be managed, and therefore it provides the foundations for the leading role and stabilizing effect of the empirical element. On the other hand, not only do we see genuine theoretical development but also there appears to be a way to go back and find redefinable previous steps which keep their relevance up to a certain, well defined extent. This does not mean that all previously effected steps are found and acknowledged, but that there seems to exist a selected sequence of steps which can be reobtained *within some redefinition*, such that the older concepts become recognizable from the point of view of the new ones. (They are not identical with, but they can be approached from, the new ones: for example, classical observations in quantum theories or Newton's laws in relativistic mechanics – see the remarks in Sect. 3.[27]) Judging also from the developments of the last one or two centuries, we can assume the hypothesis of robust scientific progress based on the symbolic character of physical knowledge and realized through a wide spectrum of theories and models. The conceptual structure promoted in this way exhibits both flexibility (in dealing with specific phenomena) and the tendency to build up coherent hierarchies (in unifying or relating various classes of phenomena). What the meanings of the open questions are and how this process is going to proceed cannot be known, of course, but a sound attitude seems to remain Helmholtz's incentive: "Hier gilt nur der eine Rat: Vertraue und handle! – das Unzulängliche wird dann Ereignis" (see Footnote 19).

*

Many of the arguments touched on in this article are discussed precisely and carefully in the philosophy of science studies – see, for instance, other chapters in this book. The discussion provided here was not intended to prove or disprove these arguments, but to illustrate the way in which they

---

[27] One can see this, also pragmatically, from the physics courses: one does not teach flogiston theory or Lorentzian electrodynamics – unless as a side example for historical reasons – but one teaches Maxwell electrodynamics and non-relativistic mechanics regularly.

are involved in practical physical thinking. As stressed at the beginning, the dialogue with philosophy is not seen as a Procustean bed (for either one) but as reciprocal questioning concerning the status and character of our knowledge. The symbolic stance is an adequate framework for setting up this dialogue.

References

J.L. Aronson, R. Harré and E.C. Way (Eds.) (1995), *Realism Rescued*, Open Court, Peru (Illinois).

M. Beller (1999), *Quantum Dialog*, The Univ. of Chicago Press.

J.D. Bjorken and S.D. Drell (1965), *Relativistic Quantum Fields*, McGraw-Hill, Inc.

E. Cassirer (1921), "Zur Einsteinschen Relativitätstheorie", in *Zur Modernen Physik*, Wissenschaftliche Buchgesellschaft, Darmstadt 1987.

E. Cassirer (1935), "Determinismus und Indeterminismus in der modernen Physik", in *Zur Modernen Physik*, Wissenschaftliche Buchgesellschaft, Darmstadt 1987.

J.T. Cushing (1998), *Philosophical Concepts in Physics*, Cambridge University Press, Cambridge.

B. d'Espagnat (1995), *Veiled Reality*, Addison-Wesley, Reading.

P.A.M. Dirac (1963), "The evolution of the physicist's picture of nature", Sci. Am. 208, p. 45.

H.G. Dosch (1997), "The Concept of Sign and Symbol in the Work of Hermann Helmholtz and Heinrich Hertz" Etud. Lett. 1-2: 74-61.

D. Giulini et al. (Eds.) (1996) *Decoherence and the Appearance of a Classical World in Quantum Theory*, Springer (Heidelberg, New York).

W. Heisenberg (1969), *Der Teil und das Ganze*, Piper and Co Verlag, Munich.

H. Hertz (1894), *Die Prinzipien der Mechanik*, Leipzig.

Th. Kuhn (1962), *The structure of scientific revolutions*, University of Chicago Press, Chicago.

Th. Kuhn (1978), *Black Body theory and the quantum discontinuity*, Oxford Univ. Press, New York.

A.I. Miller (1996), *Insights of Genius - Imagery and Creativity in Science and Art*, Copernicus/Springer, New York.

D. Papineau (Ed.) (1996) *The Philosophy of Science*, Oxford University Press, New York.

C.S.S. Peirce (1890), notes concerning the "Architecture of Theories", MS 956, in *Charles S. Peirce. Naturordnung und Zeichenprozess"*, H. Pape ed., Suhrkamp, Frankfurt am Main 1991.

K. Popper (1972), *Objective Knowledge*, The Clarendon Press, Oxford.

W. Quine (1969) *Ontological Relativity and Other Essays*, Columbia, New York 1969.

Z. Radman (Ed.) (1995), *From a Metaphorical point of View*, , de Gruyter, Berlin.

H. Reichenbach (1965), *The Theory of Relativity and A Priori Knowledge*, Univ. of California Press Berkeley, p. 73.

H. Reichenbach (1978) *Philosophie der Raum-Zeit Lehre*, Walter de Gruyter Berlin, Leipzig.

R. Rorty (1998), "Truth and Progress" in *Philosophical papers* Vol. 3, Cambridge University Press, p. 19.

S. Sambursky (1975), "Der Weg der Physik", Artemis, Zürich, München.

I.-O. Stamatescu (1994), "On Renormalization in Quantum Field Theory and the Structure of Space–Time" in *Philosophy, Mathematics and Modern Physics – A Dialogue*, E. Rudolph and I.-O. Stamatescu (Eds.), Springer, Heidelberg, New York.

I.-O. Stamatescu (1995), "Anschauung und wissenschaftliche Erkenntnis" in *Kulturkritik nach Ernst Cassirer*, E. Rudolph and B.-O. Küppers (Eds.), Felix Meiner Verlag.

I.-O. Stamatescu (1998), "Reconstruction of the Space-Time Continuum in Elementary Particle Theory", Acta Cosm. XXIV-1, 15.

G. Vollmer (1990), *Evolutionäre Erkenntnistheorie*, S. Hirzel, Stuttgart.

H. von Helmholtz (1878), "Die Tatsachen der Wahrnehmung", Rektoratsrede, in *Philosophische Vorträge und Aufsätze* Akademie-Verlag, Berlin 1971.

H. von Helmholtz (1896), *Handbuch der Physiologischen Optik*, 2. Aufl., II.

S. Weinberg (1980), "Conceptual Foundations of the Unified Theory of Weak and Electromagnetic Interactions", *Rev. of Mod. Phys.* 52, p. 515.

S. Weinberg (1993), *Dreams of a Final Theory*, Pantheon, New York.

H. Weyl (1928), *Die Philosophie der Mathematik und der Naturwissenschaften*, Oldenbourg, München 1976.

W.H. Zurek (1991), "Decoherence and the Transition from Quantum to Classical", *Phys. Today* 44, p. 36.

Part II

Views on Symbol
in the Philosophy of Science

3. The Symbol in the Theory of Science: Duhem's Alleged Instrumentalism or Conventionalism and the Continuity of Scientific Development

Karl-Norbert Ihmig

1 Changes in the Theory Concept During the 19th Century

Pierre Duhem is certainly one of the classic contributors to the theory of science. Even today, his work *The Aim and Structure of Physical Theory*, first published in book form in Paris in 1906, continues to offer a rich reservoir of ideas and arguments that may contribute to a vitalization of contemporary discussions. On closer inspection, Duhem's arguments reveal one central point, namely, how frequently he emphasizes the symbolic nature of physical theories. In the following, I wish to consider a few questions arising from Duhem's definition of the role and function of symbols in physical theory. In the first section, I shall take a closer look at how the theory concept changed during the 19th century. The second section will then consider Duhem's concept of a theory of science against the background of these changes. This would seem to confirm the general opinion that the way he views scientific theories as systems of symbols is in line with an instrumentalist or conventionalist interpretation. The third section will examine those arguments that seem to support such an interpretation in the light of recent studies on Duhem, and show that none of them are conclusive. The fourth section will then analyze the ideal of natural classification that physical theories approach successively according to Duhem's concept of physical theory. It is shown that the epistemic status of a physical theory derived from this concept of convergence goes beyond a positivist understanding. The fifth and final section will ask how far Duhem can draw on the history of science to support his postulate on the continuity of scientific development in the sense of a convergence towards the ideal of natural classification. This issue is illustrated with the example of the historical development of Newton's theory of universal gravitation.

Traditionally, two main features are attributed to a scientific theory: that it provides us with knowledge about the world around us, and, that this knowledge is also certain. For a long time, the Aristotelian model of science and the attempts to combine it with Euclidean geometry was held to be the paradigm that would unite these features.[1] This resulted in a system of axioms, postulates and definitions, confirmed theorems and corollaries that developed over time into the epitomical model of scientism (Schüling, 1969, or Arndt, 1971). This ideal of science also retained its validity into modern times when many of Aristotle's other propositions, particularly those concerning his natural philosophy, were no longer considered convincing.[2] The soundness of the scientific knowledge in line with this ideal was due not only to the certainty regarding the fundamental principles and axioms, but also to the necessity and universality of the methodological conclusions applied in it. The relation to reality consisted, first, in the evidence for the assumed principles and, second, in the way these were anchored in a comprehensive philosophical or metaphysical system. It was almost impossible to conceive of an isolated scientific theory outside such a system. Descartes, for example, illustrated this relation with his image of the 'tree of philosophy', whose roots are metaphysics, whose trunk is the universal laws of nature and cosmology and whose branches compose all the other sciences (Descartes, 1955, Letter to Picot, p. XLI–XLII).

Compared to this picture, the concept of a scientific theory went through several fundamental changes in the second half of the 19th century that emerge from a series of considerations in epistemology and the theory of science. Most of them arose in the discussions on methodological principles that took place within the sciences during the 19th century. As far as the natural sciences are concerned, one can mention, for example, the critical discussion on the concept of substance and causality. Hermann von Helmholtz and Ernst Mach called for these concepts to be replaced by the concepts of law or function. Both scientists also invested much effort in reformulating the nature of scientific concepts, focusing particularly on their *symbolic* or *representative* nature (Ihmig, 1993). In mathematics, the development of projective and non-Euclidean geometries led to doubts about the role of intuition as a source of proof for mathematical theorems (Volkert, 1986). This was joined by a reinterpretation of the axiom concept that was essentially the work of Hilbert. Mathematical axioms were no longer conceived as intuitively evident propositions, but far more as implicit definitions of the fundamental concepts arising in a system of axioms (Hilbert 1972, p. 2-4). Furthermore, increased use of the group concept within geometry and other domains of mathematics was linked to a change in the object concept in geometry or the concept of geometry in general. The actual objects of geometric study were no longer

---

[1] An assimilation of the Aristotelian concept of science as developed in the *Posterior Analytics* to the geometric methods of Euclid can be found in Euclid's commentator Proclus (Fritz, 1971).

[2] An example of this is Galileo's concept of science that still showed a strong orientation towards Aristotle's ideas, even though Galileo no longer accepted the fundamental principles of Aristotelian natural philosophy (McMullin, 1978).

isolated, concrete figures given in intuition, but the *operations* by which they are transformed into each other (Klein, 1974).

The outcome was an understanding of theory that focused not only on the symbolic nature of scientific theories but also on their independence from metaphysical assumptions. Within the natural sciences, this was expressed in Heinrich Hertz's *Prinzipien der Mechanik* (Principles of mechanics) published in 1894. Hertz points out explicitly that scientific theories are mere pictures or models that cannot claim any direct relation to reality. A direct representational relation between theory and reality is replaced by the conception of a *structural* correspondence or analogy between the two domains. This assumes that the reason - consequence relations that can be developed within a theory possess some correspondence to the cause - effect relations in real phenomena. Hertz expresses this as follows: "We construct internal simulacra or symbols for external objects, and we construct these in such a way that the consequences of the images that require consideration are always once more the images of the necessary consequences in nature of the objects represented".[3] This concedes, in principle, the possible existence of several such models that may reproduce the real cause-effect relations equally well within their theoretical frameworks. Hertz himself developed three such models for mechanics. Whereas the Newtonian model of mechanics was based on the four fundamental concepts of space, time, mass and force, and the energetic system of mechanics on the principles of space, time, mass and energy, Hertz's system recognizes only three fundamental concepts, namely, space, time and mass. Considerations that take a similar direction can also be found in the work of Hilbert when he views an axiomatic theory as a "framework or scheme of concepts along with their necessary relations to each other" that can be applied "to an infinite number of systems of basic elements" (Frege, 1976, p. 67). A theory in this sense is no longer conceived as a context for internal proofs with regard to the content of a particular science, but far more as a form for a possible proof context that, in principle, always allows for several possible interpretations or models.

2   The Symbolic Nature of Scientific Theories

It would also seem to be quite easy to classify Duhem to the line of development sketched here when his concept of a physical theory is inspected more closely. Duhem defines a physical theory as "a system of mathematical propositions, deduced from a small number of principles, which aim to represent as simply, as completely, and as exactly as possible a set of experimental laws."

---

[3] "Wir machen uns innere Scheinbilder oder Symbole der äußeren Gegenstände, und zwar machen wir sie von solcher Art, daß die denknotwendigen Folgen der Bilder stets wieder die Bilder seien von den naturnotwendigen Folgen der abgebildeten Gegenstände" (Hertz, 1910, p. 1).

(Duhem, 1962, p. 19). In this context, he points to four operations that are essential for its constitution. I shall now explain these in more detail.

The first step is to label certain physical properties as the most simple or primary qualities and assign them mathematical symbols, that is, numbers, magnitudes, and so forth. This involves a transformation or translation process that changes "a concrete" or "a practical fact" into "a theoretical fact" (Duhem, 1962, p. 133). This transformation has the character of a classification of a sign, the numerical (or geometric) symbol, to the signed, namely, the physical property. There is no need for any natural relation between sign and signed. Therefore, the translation process permits a certain freedom regarding the type of classification. At the same time, a property or a phenomenon becomes objectivized through this transformation, because it now becomes possible to establish an exact relation to other properties or phenomena and to describe the resulting interdependencies in mathematical terms. However, the translation of a practical fact into a theoretical fact cannot always be performed unequivocally. It has to be anticipated that *one* practical fact may be translated into *several* theoretical facts. As far as the selection of simple or primary properties is concerned, Duhem points out that these attributes can only be applied in a relative sense. This is not a distinction within the framework of a metaphysical system, but refers to the practical implementation of the analysis or reduction of complex phenomena; an analysis that depends essentially on what methods and means are currently available. Should these change or improve, then it may well become possible for properties that were previously seen as primary qualities to lose this status. Hence, such a distinction is always relative. Furthermore, a categorical transition becomes linked to the translation process as soon as a quality is changed into a quantity. The symbol and the symbolized, sign and signed, belong to completely different categorical levels.

Duhem cites the formulation of hypotheses as the second operation involved in the formation of theories. This consists in a combination of the numbers, sizes or geometric figures determined in the first step. The outcome of these combinations can be principles, basic laws, basic equations or possibly even basic axioms and definitions that are all summarized under the concept of *hypotheses*. He calls them "hypotheses" because, although they form the foundations of the theory, they bear no direct correspondence to actual physical facts. This means that their formulation is also not determined by such facts, but that the physicist has a free choice restricted only by the need for internal consistency within the theory in question. Thus, at this point, one finds oneself in a realm of symbols that, in turn, are combined with new symbols. The hypotheses make no claim to contain assumptions about the essence or nature of the material things (Duhem, 1962, p. 219).

Once the foundations of a theory are established in line with this second step, then its mathematical expansion follows in the form of a deductive derivation of consequences. The deductive steps proceed without any need to refer to physical facts. It is not necessary to test the legitimacy of each step itself by comparing it with reality. As a result, the second and third stage of theory formulation possess no direct relation to reality.

This does not come about until the fourth step which provides for a comparison of the consequences derived from the theory with the experimental laws that should represent or classify this theory. This calls for a further transition that makes it necessary to leave the symbolic level and translate the theoretical statements into statements on the physical properties of bodies. However, such translations are not always unproblematic. If measurements are too imprecise, it may well be found that *one* concrete fact corresponds to *several* theoretical facts. This means that the outcome of the deduction may also correspond to not just one but to several theoretical facts. To compare this outcome with the outcome of the experiment, it is necessary for the retranslation to test whether this bundle of theoretical facts corresponds exactly to one practical fact. If an unequivocal translation is possible, two paths become available. Either the measured results agree with those predicted by the theory within the bounds of error, in which case, the theory is confirmed, or they do not agree, and the theory has to be changed or rejected.

At least five general features of Duhem's symbol concept can be derived from these considerations. First, symbols are signs that describe or represent something. Initially, it does not matter whether they represent things, properties or structures. Second, there is no need for a natural relation between sign and signed. For example, we may view rising smoke as a sign of fire, and, in this case, both parts of the relationship are located on the same phenomenal level and linked together by naturally occurring processes. However, this is specifically not the case with Duhem's symbols. They belong to another level or to another category than the things they symbolize. Third, symbols never occur in isolation. They are always part of a symbol *system*, in which the focus is not so much the signs themselves but far more the relations between signs that are valid within the system and their laws of combination. What is decisive for Duhem is the possibility of drawing conclusions from the system after fundamental principles or hypotheses have been distinguished. Fourth, symbols are inherently neutral with regard to the assignment of truth values. A symbol can be neither true nor false, but only appropriate or inappropriate. Its selection can be expedient or inexpedient for depicting a condition. "Applied to a symbol," according to Duhem, "the words 'truth' and 'error' no longer have any meaning." (Duhem, 1962, p. 168). Finally, a further feature of Duhem's symbol concept is that the type of relation between the general and the particular that the symbol expresses differs from the abstract relation to be found in the general concepts of everyday experience.[4]

---

[4] Whereas generalizations of everyday experience are based on abstractions "that emerge spontaneously from concrete reality" (p. 167), the symbolic generalizations of physical theories require a complex process of mediation that may go on for several years (Duhem, 1962, p. 165- 168).

3   Duhem: An Instrumentalist or a Conventionalist?

Duhem's claims regarding the symbolic nature of theories and his attendant holism postulate would seem to indicate an instrumentalist or conventionalist conception of physical theory, because, despite the postulated agreement between theory and experience, a theory itself may be formulated largely independently from this data material, and, moreover, it is not specified explicitly by experimental findings but always permits certain possibilities of choice.[5] A series of further indices can be named in support of such an interpretation. These include, for example, his sceptical attitude towards the explanatory claims of scientific theories. In his eyes, it is not the explanation, but the *representation* of experimental laws that should be the goal of a physical theory. This is due to the demand for the autonomy of physics that must not be allowed to become dependent on metaphysical theories. Theories should not provide explanations, because *explanations* assume a metaphysical distinction between the phenomenon itself and a reality that it veils.[6] A further indication is his rejection of a mechanist or atomistic explanation of nature that seems to be a consequence of his rejection of the explanatory claims of physical theory. This is because such ways of explaining nature contain assumptions on how the essence or the nature of the reality behind the phenomena is composed, and they use these assumptions to try to explain the visible phenomenal world. One indication for a conventionalist attitude in Duhem is considered to be the way his holistic interpretation of the relation between physical theories and reality follows similar postulates to those taken by Poincaré, who is viewed as the founder of conventionalism. Finally, the affinity between Duhem's conception and Ernst Mach's positivism is pointed out. This is founded particularly on their common belief that the goal of scientific theories is the economic representation or description of phenomena.[7]

---

[5] This view, which can be traced back presumably to Popper, has been adopted by a number of modern commentators. See, for example, Worrall (1982); Dolby (1984); Christie (1994); Schäfer (1978), introduction, p. XIII, p. XXI. Diederich presents a more cautious judgement when noting that Duhem's holistic conception of the relation of physical theories to experience "may permit conventionalist strategies" (Diederich, 1974, p. 69).

[6] Under "explanation", Duhem understands the following: "To explain (explicate, explicare) is to strip reality of the appearances covering it like a veil, in order to see the bare reality itself." (Duhem, 1962, p. 7).

[7] Duhem emphasizes this agreement himself (Duhem, 1962, p. 21-22). In his paper "Physique de Croyant" published in 1905/1906, he writes: "we understood that physical theory is neither a metaphysical explanation nor a set of general laws whose truth is established by experiment and induction; that it is an artificial construction manufactured with the aid of mathematical magnitudes; that the relation of these magnitudes to the abstract notions emergent from experiment is simply that relation which signs have to the things signified; that this theory constitutes a kind of synoptic painting or schematic sketch suited to summa-

Despite these indications that Duhem held an instrumentalist or conventionalist conception of physical theories, which seem to be based on his claims regarding their symbolic character, such a classification has started to be questioned in recent times.[8] To a large extent, these doubts have been motivated by the problem of combining Duhem's conception of theory with his comprehensive research on the history of science. Although this link is also found in Ernst Mach, whose positivist attitude is beyond doubt, it possesses another status in Duhem. This is because the critical potential that the history of science makes available for meta-theoretical claims regarding the essence of physical theories specifically *fails* to confirm a positivist understanding of theory in Duhem's work.

The recent studies have also shown that a closer inspection of the above-mentioned indications, which seem to speak so convincingly in favour of a positivist conception of theory in Duhem, reveals that it does not fit so well into the body of Duhem's work. The exact meaning of Duhem's claim that a physical theory should not provide an explanation but only a representation or description of the phenomena depends on which concept of explanation he has in mind. He considers that explaining a phenomenon means to study the reality it veils and to derive the phenomenon in question from this. Because assumptions about such a veiled reality cannot be justified within physics itself, physics becomes dependent on metaphysics if it adopts them. Duhem uses the phenomenon of magnetism to illustrate how explanations of a phenomenon may vary according to different metaphysical assumptions about the nature and essence of bodies.[9] He works through the assumptions underlying the explanations of magnetism in Aristotle, Boscovich, the atomists and the Cartesians. The outcome is a multitude of incompatible conceptions. These incompatibilities arise from the contradictions between the various metaphysical systems. Hence, the problem is that when physics is dependent on metaphysics, these contradictions are transferred to the domain of physics without the slightest opportunity to decide whether these different metaphysical conceptions are correct on the basis of the methods and procedures within physics itself. This is because none of the metaphysical systems are able to derive their own detailed principles and laws that could be subjected to a scientific test.[10] This reveals the crucial point in Duhem's criticism of the explanatory claims of physical theories. Under all circumstances, physics must

---

rize and classify the laws of observation." (Duhem, 1962, p. 273–311; quotation from p. 277 [originally published as P. Duhem: Physique de Croyant, Annales de Philosophie Chrétienne 54, 1905/06, p. 44-67, p. 133-159]. He adds: "Our interpretation of physical theory is, therefore, essentially positivist in its origins." (Duhem, 1962, p. 279).

[8] Maiocchi, 1985; Martin, 1991; Needham, 1998.

[9] Duhem, 1962, p. 11-13.

[10] "Now, no metaphysics gives instruction exact enough or detailed enough to make it possible to derive all the elements of a physical theory from it." (Duhem, 1962, p. 16).

avoid being based on dogmatic and a priori assumptions about the essence and the nature of the things from which the elements of a physical theory cannot be derived and which, as a result, also *evade* direct examination through experience. This attempt to keep physics out of the field of metaphysical controversies is, nonetheless, indifferent towards either a realistic or a positivist interpretation of physical theories in general. Then, even for a realist, the warning against dogmatic assumptions that cannot be justified in physical terms would not fall on deaf ears.

Similar statements can be made about his criticism of atomism. Nowhere in his work does he claim that atomist theory is inadequate empirically in the sense of instrumentalist claims because atoms cannot be perceived directly (Needham, 1998, p. 35-36). Instead, his scepticism regarding the existence of atoms and molecules is based on the consideration that atomism particularly due to its tendency to view the presentation of a phenomenon through a mechanical model as an explanation has the greatest difficulty in integrating new discoveries and details into its theory and is obliged continuously to fall back on ad hoc modifications in the effort to do this.[11] This objection can also be understood as meaning that the ontological and metaphysical assumptions of atomism are not only dogmatic but also too indeterminate to serve as a suitable basis for deriving the experimental laws of a scientific theory (in this case, chemistry). Moreover, further arguments against the atomist interpretation of chemical structures can be found that permit absolutely no conclusions in favour of instrumentalism or positivism (compare Needham, 1998, p. 50-55). Duhem's analysis of atomism provides absolutely no conclusive indicators that his conception of theory has an instrumentalist foundation.

A further objection to an instrumentalist position in Duhem is that this would force him to deny *in principle* that a scientific theory is able to make true statements about nature. He would be obliged to limit the purpose of a theory to its function of providing an economic summary of observational data. One premise of instrumentalism is the existence of observational data that are independent of theory and may provide a criterion for the selection of theories. However, Duhem considers that it is highly questionable to assume a theory-free basis for observational data, because all experimentally obtained findings also depend on theoretical assumptions, particularly those that have to be assumed for the design and implementation of an experiment. Every experiment, as long as it serves to test a theory, already takes the acceptance of a series of theoretical assumptions for granted, starting with those underlying the construction of the experimental equipment. Without assuming

---

[11] "Then, as the experimenter's discoveries become more numerous and detailed, he will see the atomist's combinations get complicated, disturbed, overburdened with arbitrary complications without succeeding, however, in rendering a precise account of the new laws or in connecting them solidly to the old laws." (Duhem, 1962, p. 304).

the correctness of these theories, it would be "impossible to regulate a single instrument or to interpret a single reading." (Duhem, 1962, p. 182). If an inconsistency arises between theoretical prediction and experimental result, then the only possible conclusion is that a modification needs to be carried out within one theory or group of theories. Where this has to occur or *which* hypothesis has to be modified is not specified at all. In particular, no un-equivocal experimental decision can be made between two hypotheses in the sense of an *experimentum crucis*. This is because the mathematical schema of indirect proof cannot be transferred to physics. Confirming the falsity of one hypothesis is no necessary confirmation of the correctness of the other as long as there is still, in principle, a series of further conceivable hypotheses. Moreover, Duhem's work contains no explicit discussion of the problem of how to proceed when several theories provide an equally good presentation of the empirical data. Which theory should be preferred? It seems that this problem, which is crucial for an instrumentalist, does not arise in this form for Duhem. The reason for this becomes clearer when one takes a closer look at Duhem's conception of natural classification.

The postulate of the holistic relation to experience of scientific theories sketched above has led many commentators to mention Duhem in the same breath as Poincaré's conventionalism. Poincaré discriminates, and Duhem as well, three levels in a physical theory: facts, experimental laws and principles. The principles, which form the peak of this hierarchy, are no longer subject to control through experience in Poincaré's conception. They have definitory character and, therefore, cannot be rejected empirically. Poincaré cites Newton's axioms of motion as examples of such principles (Poincaré, 1914, p. 91-107). Therefore, when Duhem considers that a principle cannot be rejected empirically as an element of a theory, he seems to be wanting to make the same claim as Poincaré. Referring explicitly to the principle of inertia and its interpretation by Poincaré, Duhem writes: "We cannot, therefore, attempt an experimental verification of the principle of inertia." (Duhem, 1962, p. 213). Nonetheless, Duhem continues, it would be a fallacy to believe that this places such principles or hypotheses beyond experimental control in principle. It is only in *isolation* that they cannot be tested empirically. As an element of a theory, they may well be rejected *indirectly* as a result of empirical findings: "Taken in isolation these different hypotheses have no experimental meaning; there can be no question of either confirming or contradicting them by experiment. But these hypotheses enter as essential foundations into the construction of certain theories [...]. The object of these theories is to represent experimental laws; they are schematisms intended essentially to be compared with facts." (Duhem, 1962, p. 215-216). Just as one is hardly compelled to dismiss a certain hypothesis or a specific principle in a theory because of discrepancies between theory and experience, one is not obliged to raise it to a definition and thus immunize it from any empirical test. Hence, Duhem's holistic conception does not lead conclusively to conventionalism.

Therefore, one more issue still needs to be explained; namely, Duhem's relationship to the positivism of Ernst Mach. This seems to be a serious issue, because Duhem maintains explicitly that his concept of theory is "essentially positivist in its origins". However, here as well, things are not as explicit as they seem. First, the claim that a theory conception has a positivist *origin* is not equivalent to saying that it is itself positivist. The reference to the origin may also be understood as meaning that Duhem is thinking of a *development* of the theory concept that starts out with a positivist concept but in no way stops there. A closer inspection of the second chapter of the first part of *The Aim and Structure of Physical Theory* seems to confirm this suspicion. It reveals a sequence of three definitions of the purpose of a theory that are certainly based on the positivist theory concept, but cannot *all* be explicated as a consequence of this concept. Duhem assumes, exactly like Poincaré, that a theory can be structured into the different levels of "facts", "experimental laws" and "principles". This is a hierarchy of stages of universality, because laws proceed from the abstraction of a set of concrete facts, and principles summarize a number of laws in a similar way. Accordingly, the purpose of a theory seems to consist in taking an otherwise scarcely comprehensible set of individual facts and replacing them in the most economically way possible through simple laws, and replacing a multitude of laws through principles. A theory would then be nothing other than a system of signs representing a certain set of facts that belong to a specific field of research. By emphasizing the economic utility of a theory, Duhem refers to Ernst Mach, whom he cites explicitly in this context (Duhem, 1962, p. 21-22). Hence, the purpose of a theory presented here is to obtain an *overview* of a certain field quickly and easily. However, it is necessary to go beyond this definition of purpose, because: "Theory is not solely an economical representation of experimental laws; it is also a *classification* of these laws." (Duhem, 1962, p. 23). This means that the mere description of a multitude of laws through one unifying sign should be understood as only a first approach to, but no final definition of, the purpose of a theory. A classification of these laws is sought as well. Hence, laws are combined into groups that show a degree of relationship. However, in the framework of a scientific theory, this relationship is not based on superficial similarities or analogies, but on the logical relation of the laws to the principles. In optics, for example, the various principles of refraction, interference and diffraction lead to the corresponding experimental laws being combined into various classes that are assigned to each of the respective principles. The principle of refraction groups the laws that define the colour spectrum generated by a prism along with the laws of the colours of the rainbow, but it excludes the laws responsible for the phenomenon of Newton's rings and assigns these to another principle. Beyond providing a mere overview, classification establishes an *order* between the experimental laws in terms of the principles of a theory. The utility of this order is that it makes the laws easier to *apply*. Duhem compares this classification with a toolbox in which instruments that serve different tasks are placed in different compartments. This makes them easier to handle than if they were thrown into one compartment indiscriminately.

However, Duhem does not stop at this description of the purpose of a theory either. A physical theory does not just strive towards a classification of the laws, but also towards a natural classification: "the more complete it becomes, the more we apprehend that the logical order in which theory orders experimental laws is the reflection of an ontological order, the more we suspect that the relations it establishes among the data of observation correspond to real relations among things, and the more we feel that theory tends to be a natural classification." (Duhem, 1962, p. 26-27). Thus we find a progression of definitions of the purpose of a physical theory starting with the positivist conception of the economic abbreviation for a set of facts, going on to the perspective of classifying experimental laws, and finishing up with the idea of natural classification. It may still be possible to justify the first two definitions of purpose within the framework of a positivist theory conception. However, this would become difficult when the goal is a natural classification, because such a goal involves the connection of theoretical considerations and ontological presuppositions that lies beyond the horizon of a positivist approach. Hence, Duhem's reference to Mach also does not permit any unequivocal conclusion that Duhem's theory concept is positivist, because this cannot be reconciled with the last-mentioned definition of purpose. To understand why this is so, it is necessary to take a closer look at the origin, the concept and the reach of the purpose of a "natural classification" in Duhem's work.

4   Natural Classification and the Systems Concept

The first time Duhem formulates the goal of the natural classification of a physical theory is in his article *L'école Anglaise et les Théories Physiques, à propos d'un livre récent de W. Thomson* published in 1893.[12] Its contents include Duhem's reaction to the engineer Eugène Vicaire's criticism of his paper *Quelques Réflexions au sujet des Théories Physiques* published the year before. This article was Duhem's first publication on a topic in the philosophy of science. He used it to develop an instrumentalist concept of theory completely in line with the contemporary positivist philosophy of science expounded by August Comte. In particular, he associated the symbolic character of a theory exclusively with its ability, by dint of the symbols it uses, to represent a set of facts. Accordingly, the primary purpose of a physical theory was economy of thought in the sense of Mach. Vicaire raised several objections to this

---

[12] This appeared in Rev Quest Sci 24, 1893, p. 345-378. On the following, compare Martin, 1991, p. 29-33.

understanding of theory, including criticisms of the principle of economy of thought. If the main purpose of a theory were to offer only a memory aid for a multitude of facts and laws, then its mathematical form would have to be viewed as superfluous. It would suffice to replace it with a collection of mnemotechnic signs each representing a set of facts of experimental laws. In replying to Vicaire's criticism, Duhem admitted expressly that the principle of economy of thought provided only an inadequate description of the essence of a physical theory. It was necessary to seek a systematic relationship between the principles, laws and facts presented in the theory so that it does not just provide a description of the phenomena but makes it possible to understand their *unity* as well. However, at the same time, he pointed out that this goal of unifying the phenomena represented an ideal that could not be justified by dint of the experimental methods of physics, but can only be postulated. He described this ideal as a "natural classification".

Duhem took the concept of natural classification from biology. In biology, a classification generally means a division of a multitude of concrete individuals into genera, species, subspecies and so forth. This division is based on a procedure of abstraction during the course of which, by retaining general features and ignoring differences that are considered unessential, it is possible to move from complex, concrete facts to increasingly simple and general determinations. The resulting classification is then a hierarchy of concepts divided into stages with different levels of generality. Such an ideal schema represents a *natural* classification when the relationships and dependencies it indicates correspond to real relations between the concrete living things. For example, a certain classification of vertebrates indicates a theory of evolution, so that the question whether this has led to a natural classification can be tested in physiology and palaeontology. The transfer of the concept of natural classification from biology to physics is therefore conveyed by an analogy. Duhem considers that an unmistakable indication that physical theories also strive towards a natural classification is their ability to make experimental predictions on the occurrence of phenomena under certain conditions or the fact that any scientific progress is possible at all. "If, [...], we recognize in the theory a natural classification, if we feel that its principles express profound and real relations among things, we shall not be surprised to see its consequences anticipating experience and stimulating the discovery of new laws." (Duhem, 1962, p. 28).

Which property of a theory forms the basis for such predictions? In his paper *Physique de Croyant*, Duhem uses an interesting comparison. Assume that a man is collecting molluscs, and he orders the molluscs he finds according to their colour in different drawers. Assume further that the collector has not anticipated any need for a drawer for blue molluscs. Can we now assume that blue molluscs do not exist? Alternatively, the collector has prepared a drawer for blue molluscs, but not found any so far. Can this lead us to anticipate that blue molluscs must also exist in reality? In both cases, according to Duhem, we would laugh at such anticipations and conclusions (Duhem, 1962, p. 297-298). Why are things different with a physical theory? Why does it seem justified to assume that the experimental results predicted by the theory will also be confirmed here? "Obviously because the classification of this collector is a purely arbitrary system not taking into account the real affinities among the various groups of mollusks, whereas in the physicist's theory there is something like a transparent reflection of an ontological order." (Duhem, 1962, p. 298). A physical theory is not a mechanical aggregate of individual components, but a system whose regularities indicate an organic relationship.[13] Because of this systematic character, it not only goes beyond merely establishing and gathering facts or experimental findings, but also offers an opportunity to approach reality, to approach the real ontological order. If the goal of a theory is only to provide an economic abbreviation for an otherwise difficult to grasp set of facts as well as an arbitrary arrangement and classification of the same, then physicists would be in no better position than our mollusc collector. They could only make statements about the *actual*, but not about the *potential*. Statements about *potential* dependencies and relations can never refer directly to the "actual" but are an additional product of our intellect.

The goal of the systematic unity of theory that links up with the ideal of the natural classification can therefore *not* be justified by dint of positivist experimental principles.[14] Positivists should have no difficulty in representing one domain of phenomena through a certain group of experimental laws and another domain through a further group, regardless of whether these groups and the principles assigned to them are compatible. Physicists, in contrast, strive towards a *unification* of these domains without being able to confirm this effort in any way at all on the basis of a positivist concept of science.[15] However, does this not mean that physics finally has to fall back on meta-

---

[13] See Duhem, 1962, p. 32, where he talks about a "fully formed organism", or p. 187, where he states: "Physical science is a system that must be taken as a whole; it is an organism [...]".

[14] Duhem, 1962, p. 298: "To the extent that physical theory makes progress, it becomes more and more similar to a natural classification which is its ideal end. Physical method is powerless to prove this assertion is warranted, but if it were not, the tendency which directs the development of physics would remain incomprehensible. Thus, in order to find the title to establish its legitimacy, physical theory has to demand it of metaphysics."

[15] Duhem, 1962, p. 219: "Now, we have recognized that it is impossible to construct a theory by purely inductive method. [...] Therefore, we shall not be averse to admitting among the fundamental bases of our physics postulates not furnished by experiment." See also p. 293: "No scientific method carries in itself its full and entire justification; it cannot through its principles alone explain all these principles. We should therefore not be astonished that theoretic physics rests on postulates which can be authorized only by reasons foreign to physics. Among a number of these postulates is the following ones: Physical theory has to try to

physical principles in order to achieve its own ideal? If the goal of a physical theory is a natural classification, and this goal is equivalent to striving towards a systematic unity of all theories, because the potency of its prognoses and thus of scientific *progress* depends on this, then drawing on metaphysical principles is one of the conditions that make it possible for a theory to progress. Does this not leave Duhem in flagrant contradiction to his initially proclaimed intention to emancipate physics from metaphysics?

An open contradiction would exist only when the epistemic status of the metaphysical or philosophical assumptions transcending physics that physics is forced to use were the same in both cases. However, this is not the case. The insight that the ideal of natural classification finally refers to principles and postulates that lie outside the framework of positivist methods needs to be understood more as the outcome of a dialectic argumentation.[16] First of all, Duhem only rejects metaphysical assumptions that contain dogmatic statements on the essence and nature of things such as the existence of atoms or the reduction to mechanistic actions. What was crucial about these claims was that they were outside experimental control *in principle*. After rejecting dogmatic metaphysics, Duhem consistently follows the path of basing physics *exclusively* on positivist methods. However, in the long run, this route also proves to be impassable; first, because of the holistic character of physical theories, and, second, because of the limitations of the inductive methods whose deficits he tried to demonstrate with the examples of Newton and Ampère. For Duhem, the conclusion that has to be drawn from the failure to subject a physical theory *one-sidedly* to the rule of metaphysics or positivist methodology is the need to search for an ideal that may link together *both* elements in an appropriate way. If it proves to be necessary to go beyond the positivist understanding of theory, then this should be done within a conceptual framework that leaves two major cornerstones of a physical theory untouched: its mathematical structure and the possibility of testing it through experience. The attempt to synthesize both elements is expressed in two operations: the selection of hypotheses and the correction of a theory to fit the facts more precisely. These operations point in two different directions. When selecting hypotheses, the focus is on the goal of unification in the sense of striving towards a natural classification. One is seeking simplification and generalization. Corrections, in contrast, make theories increasingly more complex. They focus attention on the particular, the individual: "The physicist who complicates the theoretical representation of the observed facts by corrections, in order to permit this representation to come to closer grips with reality, is similar to the artist who, after finishing the line sketch of a drawing, adds shading in order to express better on a plane surface the profile

---

represent the whole group of natural laws by a single system all of whose parts are logically compatible with one another."

[16] Martin has drawn attention to this dialectical strategy against the background of Duhem's reading of Pascal (Martin, 1991, p. 108-111).

of the model."[17] This means that physical hypotheses or principles are not fundamentally beyond experimental control, even when their selection cannot be justified directly through experimental findings. However, they can be tested only *indirectly* when they promote an order and dependence within the constitutive elements of a theory (principles, concepts, laws, etc.) that opens up the possibility of corrections; in other words, of increasing complexity without losing the systematic context. Duhem's criticism of mechanistic models is aimed directly at the fact that such corrections can only be carried out at the price of the loss of unity. Both directions, towards the special and individual as well as towards the general, work together in the formulation of a physical theory and are mutually dependent. Taking this relation into account makes it possible for a physical theory to use metaphysical principles without entering into a one-sided dependence on assumptions from dogmatic metaphysics and, in this way, making physics *subordinate* to metaphysics.

Nonetheless, corrections are only possible up to certain limits. A complete match between theory and reality is ruled out in principle. This is because physics applies the symbolic relations of mathematics without its principles possessing the direct evidence of mathematical axioms. Duhem emphasises that, for example, the principles of Newtonian dynamics are far removed from the insights of common sense.[18] He strongly criticizes Euler's attempt to define the basic concepts of mechanics by referring to ideas from daily life (Duhem, 1962, p. 261–263). In contrast to geometry, in which certainty and truth are associated with the simplicity of its objects, physics handles concrete, complex objects whose theoretical exploration requires a tiresome and difficult process of analysis. This reveals, according to Duhem, a remarkable "relation of indeterminacy". The less they are analysed and the more imprecise ideas are, the closer they are to common sense, and this means the more they are certain and true. However, the more precisely one analyses and orders them, the more abstract and uncertain they become. Although the creators of mathematical physics have succeeded in taking a brilliant step forward, "nevertheless, they have not been able to make clarity and order come into physics and become fused immediately with self-evident certainty, as they have in arithmetic and geometry. All they have been able to do is to confront the multitude of laws obtained directly from observation, laws that are confused, complex, and disorderly but endowed with a certainty directly ascertainable, and to draw a symbolic representation of these laws, an admirably clear and orderly representation, but one which we can no longer even properly say is true." (Duhem, 1962, p. 266). Physics resides, unlike mathematics, in the intermediary zone between the directly observed truths

---

[17] Duhem, 1962, p. 158. See also p. 175: "The mathematical symbol forged by theory applies to reality as armor to the body of a knight clad in iron: the more complicated the armor, the more supple will the rigid metal seem to be".

[18] On the role of "common sense" in Duhem and its relation to Pascal's concepts of "connaisance commune", "bon sens" and "finesse", see Martin (1991), p. 79-90.

of common sense, which are characterized by their reliability and certainty, and abstract symbolic deductions whose truth is always problematic despite their precision, clarity and necessity. Between these two domains, we find a "continual circulation and exchange of propositions and ideas". However, because a physical theory, in accordance with the ideal of natural classification, strives towards *both* certainty and truth *and* precision and necessity, its laws, principles and hypotheses are always only provisional. If Duhem were to have embraced an instrumentalist interpretation, he would not have needed to take any interest in the first point. The provisional character that Duhem attributes to physical theories is not synonymous with their instrumentalist character.

Thus, we are finally left with one decisive question. The holistic interpretation of scientific theories, the impossibility of performing an *experimentum crucis*, the selection of hypotheses, the performance of corrections, in other words, almost all the essential elements of Duhem's philosophy of science depend on the ideal of natural classification as the goal of the formation of a physical theory. What would happen to Duhem's conception if it should prove to be a mere chimera? We can neither confirm this ideal that specifies the direction of scientific research or achieve certain knowledge of it in any other way. "Now, is it right to regard this ideal as utopian? It is up to the history of physics to answer this question; it is up to it to tell us whether men, ever since physics took on a scientific form, have exhausted themselves in vain efforts to unite into a coordinated system the innumerable laws discovered by experimenters; or else, on the other hand, whether these efforts through slow and continuous progress have contributed to fusing together pieces of theory, which were isolated at first, in order to produce an increasingly unified and ampler theory. To our mind that is the great lesson we ought to obtain when we retrace the evolution of physical doctrines." (Duhem, 1962, p. 295). This is when the systematic function of the history of science in Duhem's philosophy of science and, in particular, the continuity hypothesis become clear. Even if we are unable to anticipate any absolutely valid answer from history, it can at least provide indications of certain developmental *movements*. Any contemporary theory is always only a single snapshot in a continuous line of development. Duhem illustrates this point with an example. If we want to ascertain the target of the trajectory of a tennis ball, then it does not help us much to know its position at any one point in time. However, if we know the entire course of the trajectory that the ball has followed up to the present, then grounded ideas are possible on its further trajectory and its target. "So the history of physics lets us suspect a few traits of the ideal theory to which scientific progress tends, that is, the natural classification." (Duhem, 1962, p. 303). The final section will use a brief analysis of the history of Newton's law of gravitation as an example to discuss this postulate of Duhem.

5 Natural Classification
and the Continuity of Scientific Development

In general, according to Duhem's concept, two movements overlap in the development of physics. The one movement is characterized by a continuous coming and going of theories that dominate the science for a while before losing their meaning. "The other movement is a continual progress through which we see created across the ages a constantly more ample and more precise mathematical representation of the inanimate world disclosed to us by experiment." (Duhem, 1962, p. 306). These two movements result from the fact that, throughout their development, physical theories have always been embedded in more general metaphysical assumptions. One major precondition for the plausibility of a convergence of these theories toward the ideal of natural classification is, in Duhem's eyes, a separation of these two elements that go into the development of a physical theory. He proposes differentiating the "representative part" from the "explanatory part" of a theory, because the continuity of the development of science in the sense of convergence towards this ideal manifests solely through the development of the representative part of a theory. For example, the validity of Descartes' law of refraction does not depend on its mechanistic explanation of the light phenomena that draws on metaphysical assumptions. Descartes based his explanation on the assumption that light spreads instantaneously. After the Danish astronomer Rømer had determined the velocity of light, Descartes' explanation of his refraction law became untenable, but the law itself continued to be valid. Duhem uses this to conclude that when the explanatory part of a theory changes, the core of the representative part, which is based exclusively on principles inherent to physics (namely, experience, induction and generalization), is retained. In his own words: "Thus, by virtue of a continuous tradition, each theory passes on to the one that follows it a share of the natural classification it was able to construct, as in certain ancient games each runner handed on the lighted torch to the courier ahead of him, and this continuous tradition assures a perpetuity of life and progress for science." (Duhem, 1962, p. 32-33).

The example that Duhem discusses in most detail is Newton's theory of universal gravitation. This example should show that the selection of hypotheses, which can initially occur arbitrarily within a very far-reaching framework, becomes increasingly constrained by the historical development of a theory. What does the issue of the selection of hypotheses have to do with the convergence towards the ideal of natural classification? The ideal unites two elements, namely, the largest possible systematic unification with the highest possible precision in the representation of the phenomena. The goal of systematic unification assumes the existence of principles of unity. These can take the form of fundamental concepts, definitions, axioms, laws or general principles; in other words, they can vary in range. Duhem summarizes them under the collective term 'hypotheses' because they are not obtained by purely inductive means, and therefore, at least in the positivist sense, cannot be verified directly in empirical terms. Therefore, the unifying ability of a theory depends on the selection of principles of unity, in other words, on the selection of hypotheses. In what sense does Newton's theory come closer to the ideal of natural classification than previous theories, and which indices make this postulate plausible?

Duhem analyses Newton's gravitation theory by interpreting it as a synthesis of different hypotheses previously belonging to completely different contexts, not all in the realm of physics. In the foreground, there are three hypotheses that Duhem considers to represent the fundamental principles of the theory, namely: (a) the mutual attraction of heavenly bodies, (b) the mathematical law of attraction (the inverse-square law), and (c) the dynamic analysis of circular motion involving a force directed towards the centre and an inertia component. Essentially, Duhem limits himself to tracing the history of these three hypotheses in the various theories and contexts in which they emerged *before* Newton.[19] Regarding the hypothesis on the mutual attraction of heavenly bodies, he follows the important stations of the theory of gravity since Aristotle. He focuses on the history of explanations for the phenomenon of tides that was initially attributed to an effect of the moon in an analogue way to the effect of magnetism and later to an effect of the moon and the sun (the ideas of Morin and Roberval). Regarding the mathematical law on the propagation of attraction, he sees Newton's inverse-square law as part of a long tradition based particularly on the analogy to the law on the propagation of light. In the second half of the 17th century, this was followed by the analysis of circular motion in various components as carried out by Huygens and Hooke, an analysis that was also based on the previous work of Kepler, Descartes, Roberval and Borelli.

The history of Newton's law of gravitation involves, on the one hand, a series of completely heterogeneous reflections and unrelated theories that are assembled from various elements and draw on different sources. These include everyday experiences of gravitational phenomena; scientific measurements; Kepler's laws; Descartes' vortex theory; atomist models; Huygens' dynamics; the metaphysical theories of the Peripatetics, astrologers and physicians; as well as the use of analogies with the phenomena of light and magnetism. It would seem that this cannot form the basis for a continuity hypothesis or make the convergence of theories plausible. However, Duhem attempts to separate the three above-mentioned hypotheses, whose synthesis form the basis for Newton's theory, from their diverse contexts and investigates the evolution of each of these in its own right. Therefore, it does not seem ungrounded to assume that he regarded them as the 'representative part' that can be separated from the explanatory part. The continuity of development can then be traced back to the descriptive part by initially studying the successive development of each of these three elements with its applied field in isolation without referring to the metaphysical contexts in which each is

---

[19] See Duhem, 1962, part II, chap. VII.2., p. 220–252.

involved. Under this assumption, Newton's theory of gravitation represents a step in the direction towards the ideal of the natural classification of the phenomena, because it unifies these three hypotheses under a general principle by developing them within the framework of a mathematical system of symbols representing their order and internal cohesion. In this way, Newton succeeded not only in developing a unified theory for terrestrial and heavenly mechanics but also in adjusting the theory to the phenomena by dint of corrections, because the principle of perturbations can explain the deviation of planetary orbits from Kepler's laws.

Duhem's maxims for the reconstruction of the history of science certainly raise some problems. The first lies in the separation of the representative from the explanatory part of a theory. That which Duhem understands under the descriptive part of a theory clearly refers to the theoretical objects of physics and not to directly perceived events. According to the above-mentioned (Sect. 2) first step of the formation of a physical theory these objects depend on a transformation that is performed within presupposed (mathematical) systems of symbols. Therefore the representative part can ensure the continuity of development only under the assumption that the system of symbols applied in it is also not subject to major changes. However, there is no reason why this assumption should always be met. It is highly conceivable that, in the transition from one theory to the next, the system of symbols on which the representative part depends changes in a way that no longer permits the construction of a continuous link between the two. If, for example, one compares the symbolic representations of the fundamental laws of mechanics *before* Newton (e.g., in Kepler), in Newton himself (in the *Principia*) and in the 18th century (e.g., in Lagrange), then it is difficult to identify any "lighted torch" that each runner has handed on to the courier ahead of him.

Second, Duhem's orientation towards the goal of natural classification implies an anachronistic approach to history. His primary concern is not to compare two successive theories in terms of their total structure. He starts with the most successful contemporary theory, identifies which hypotheses are crucial to it, and then traces their historical origins while ignoring all the other assumptions with which they are linked. Therefore, he judges the development of the preceding theories exclusively from the perspective of the currently dominant theory. As a consequence, all conceptions that have not contributed directly to the breakthrough of this one theory are neglected as developmental "dead ends". As, for this reason, all side paths are ruled out in advance as objects of historical analysis, it is hardly surprising that one has to gain the impression of a continuous development towards a certain goal.

Third, and finally, it is unclear, particularly with regard to Newton's theory, how far the goal of unification should be traced back to the representative or the explanatory part. This is because Newton developed de facto not just one mathematical system of symbols from which he derived rules that he then compared with the phenomena, but he based this system on a series of ontological assumptions that are metaphysical in nature. For Newton, that which underlies the phenomena, which determines their real essence, consists in a diversity of forces acting in quite different ranges. An explanation of the "system of the world" as he intended, called for a systematic unification of this variety of forces. He was thoroughly aware that his programme for the third book of the *Principia*, in which he intended to explain the phenomena of the solar system, the gravitational fall of bodies to the earth and the tides in terms of one unified force as their common cause, was only the beginning of a much more ambitious project. He hoped that he would also be able to use the same method to trace all the other phenomena of nature back to one unified dynamic cause. Hence, it is not completely clear whether the approach towards the ideal of natural classification, in so far as it implies the goal of unification, can be traced back to the representative or explanatory part of Newton's theory.

Nonetheless, Duhem's concept of convergence draws attention to the complementarity of various methods and sources of knowledge that work together to construct a physical theory. First, his attempt to separate physical theories from both mathematics and everyday experience is important. The hypothetical or provisional status of physical theories is linked to the essence of these theories through being located in the intermediary zone that he describes so aptly, and oscillates continuously between the formulation of mathematical systems of symbols and the ideas of common sense. He does not reject the use of models in physics in principle, but points primarily to their exemplary and restricted role. This is why they cannot simply be equated with universal principles and hypotheses. Second, Duhem was right to point out that if physics strives towards the ideal of a unified comprehensive theory, it has to draw on principles and hypotheses that it can no longer justify on the basis of its own methods. The image of physics as a science depicted by positivism moves within a framework that is not broad enough to allow a conceptual grasp of the systematic progress of a theory. Progress must appear to be just as arbitrary to the positivist as to the mollusc collector who has left one drawer free for blue molluscs and then, at some time or another, find a blue mollusc. Finally, Duhem emphasizes that scientific progress in the sense of approaching the ideal of natural classification contains the complementarity of two apparently contradictory movements: the movement towards the general and the movement towards the particular. Neither of these directions can be pursued in isolation. By focusing only on the general, a physical theory soon becomes dependent on metaphysical theories. By focusing only on the particular, one would get bogged down in the search for special principles for special phenomena without considering the possibility of unifying them. It is only the simultaneous consideration of both movements that enables physics to take advantage of metaphysical principles without having to subject itself to metaphysical systems.[20]

References

Arndt, H.-W. (1971): *Methodo scientifica pertractatum. Mos geometricus und Kalkülbegriff in der philosophischen Theorienbildung des 17. und 18. Jahrhunderts*, (Berlin/New York).

Christie, M. (1994): "Philosophers versus Chemists Concerning Laws of Nature" *Stud Hist Philos Sci 25, p. 613-629.*.

Descartes, R. (1955): *Die Prinzipien der Philosophie*, transl. and comment. by A. Buchenau (Hamburg), [Amsterdam 1644].

Diederich, W. (1974): *Konventionalität in der Physik*, (Berlin).

Dolby, R. G. A. (1984): "Thermochemistry versus Thermodynamics: The Nineteenth Century Controversy" *Hist Sci 22*, p. 375-400.

Duhem, P. (1962): *The Aim and Structure of Physical Theory*, translated by P. P. Wiener (New York).

Frege, G. (1976): *Nachgelassene Schriften und wissenschaftlicher Briefwechsel*, ed. by H. Hermes/F. Kambartel/F. Kaulbach, 2. Bd.: *Wissenschaftlicher Briefwechsel*, Teil XV: Frege-Hilbert, (Hamburg).

Fritz, K. v. (1971): "Die in ΑΡΧΑΙ in der griechischen Mathematik" in Fritz, K. v.: *Grundprobleme der Geschichte der antiken Wissenschaft*, (Berlin/New York), p. 335-429.

Hertz, H. (1910): *Prinzipien der Mechanik*, [first edition 1894] Leipzig.

Hilbert, D. (1972): *Grundlagen der Geometrie*, [Leipzig 1899] (Stuttgart).

Ihmig, K.-N. (1993): "Cassirers Begriff von Objektivität im Lichte der Wissenschaftsauffassungen des ausgehenden 19. Jahrhunderts" *Philos Nat 30*, p. 29-62.

Klein, F. (1974): *Vergleichende Betrachtungen über neuere geometrische Forschungen (1872)*, Wussing, H. (Ed.), (Leipzig), p. 29-84.

Maiocchi, R. (1985): *Chimica e Filosofia, Scienza, epistemologia, storia e religione nell'opera die Pierre Duhem*, (Firenze).

Martin, R. N. D. (1991): *Pierre Duhem. Philosophy and History in the Work of a Believing Physicist*, La Salle (Illinois).

McMullin, E. (1978): "The Conception of Science in Galileo's Work", in Butts, R. E./Pitt, J. C. (Eds.): *New Perspectives on Galileo*, (Dordrecht, Boston), p. 209-257.

Needham, P. (1998): "Duhem's Physicalism" *Stud Hist Philos Sci 29*, p. 33-62.

Poincaré, H. (1914): *Wissenschaft und Hypothese*, transl. and comment. by F. and L. Lindemann (Leipzig), [Paris 1902].

Schäfer, L. (1978): "Duhems Bedeutung für die Entwicklung der Wissenschaftstheorie und ihre gegenwärtigen Probleme" in Schäfer, L. (Ed.): *Duhem, P., Ziel und Struktur der physikalischen Theorien* (Hamburg), p. IX-XLIV.

Schüling, H. (1969): *Die Geschichte der axiomatischen Methode im 16. und beginnenden 17. Jahrhundert*, (Hildesheim/New York).

---

[20] I am indebted to Jonathan Harrow for the translation of the original German text.

Volkert, T. (1986): *Die Krise der Anschauung. Eine Studie zu formalen und heuristischen Verfahren in der Mathematik seit 1850,* (Göttingen).

Worrall, J. (1982): "Scientific Realism and Scientific Change" *Philos. Q. 32*; p. 201-231.

4. Beyond Realism. Symbolism in the Philosophy of Science by Charles S. Peirce and Ernst Cassirer

Enno Rudolph

Peirce's philosophy is based upon three centers of gravity forming a system, the organization of which can be exemplified as three concentric circles. The innermost circle represents the endeavor to create an original logic of science based primarily on the history of sciences from Galileo to Mach. As the one closest to the center, this circle founds the other two; however, its scope depends on the diametes of the other two circles. The second circle represents the founding of the principle of pragmati(ci)sm.[1] It conjoins the other two circles by defining the first circle's aim of an original logic of science by drawing an unusual analogy between thought and action. The third circle represents semiotics, probably the most successful part of Peircean philosophy. While the independent reception of this field of Peirce's work is only just developing, scholars' interest in the middle field, i.e. the philosophy of pragmatism, appears to be waning.

On the whole, however, general interest in Peirce's work is steadily growing. His voluminous but scattered oeuvre includes a few selected texts which can be said to represent the three circles mentioned above. Among these texts is the key piece *How to make our ideas clear* – the core text of the Peircean system, regarded in the literature as the "birth certificate" (see Oehler, 1993, p. 23) of pragmatism, not least because the essay, which, of the several revisions, was published in 1878 as part of a series in *The Popular Science Monthly*, formulates a sort of categorical imperative of pragmatism, the so–called "pragmatic maxim". The maxim is developed gradually out of a distinctly but accessibly formulated scientific methodology in which, in more or less polemic style, he sought to distance himself from entrenched traditions in the philosophy of science, which in his day were almost exclusively European. This form of exposition makes it easier for historians of philosophy and history to determine Peirce's exact position within the history of ideas: The first part deals with the relationship of the Peircean epistemology and his logic of science in connection with the scholars, named by Peirce as his disputants, historically involved in these issues; the second part represents a discourse with Ernst Cassirer, an author relevant to the subject whom Peirce acknowledged, without, however, considering his views in detail.

---

[1] For the reason why Peirce changed the term "pragmatism" into "pragmaticism", see Peirce, 1967, p. 394.

1   Sign and Science: Peirce and His Predecessors

Peirce's historical disputants are also his adversaries: especially Descartes, Leibniz and the tradition and positions of nominalism. David Hume's influence on Peirce's reasoning is so profound that it is even reflected in his choice of words, a fact which raises the question whether, in Peirce's case, it was Hume – as rendered by Kant – who provided the "great illumination"[2] which so enduringly enlightened the latter. In fact, some years earlier, when already working on his "ideas", he admitted that "it was only through Kant's doorway that I crossed the threshold of philosophy" (see Oehler, 1993, p. 19), which can only refer to Kant in his critique of Hume.

Peirce's appraisal of Descartes and Leibniz centers on methods for defining scientific concepts. In spite of their differences, both philosophers contend that a concept can only be said to be well defined if its content is "clare et distincte". According to their view, concepts are clear only when that to which they refer is intelligible. However, Leibniz qualified this statement, noting that it was insufficient in that it did not provide the criteria with which to distinguish between different ideas. Signs, he argued, required definitions which guaranteed that the defined object cannot be mistaken for another. Accordingly, Leibniz developed a theory of concept transcending that of Descartes in that it aimed at encompassing all the characteristics of the defined object, creating, in Leibniz's words the *notion individuelle* (see Rudolph, 1989, p. 113).

However, Peirce's critique of Leibniz's classical maxim does not refer to the intention of exactitude regarding scientific terms – he was too much of an Aristotelian not to have known that both Descartes and Leibniz took on the honorable task of defining Aristotle's criteria of the *genus proximum* and the *differenzia specifica* more precisely. Peirce, however, doubts that these criteria are sufficient. His criticism emerges, albeit diplomatically veiled, in his remark that Descartes completed the transition from scholastic method of authority to that of a priori. Both Leibniz and Descartes claimed they could assess the clarity of a concept by analogy to mathematic axiom. A concept determines a priori what is attributed to the object it refers to. Concepts act as functions from which we can deduce the defined object's characteristics much as objects lined up in row. Knowing the function means *deducing* in order to define. In addition to the employment of mathematics as a model for defining concepts, philosophy states the axioms which thought cannot surpass; according to Leibniz, these axioms are expressed in the theorems of contradiction and of sufficient proof.

Peirce's main objection to these philosophies of definition emerges in the logic of his theory of "abduction", namely in its explicit role as an addition

---

[2] It was Kant who used this metaphor to illustrate the fascination he felt by discovering the relevance of permanent doubt embodied in the position of David Hume, see Kant, AA XVIII, p. 69.

to and corrective of the methods of deduction and induction. He faults these methods for preventing new developments in thought. For instance, Peirce points out that Descartes' principle of "methodic doubt", which established his reputation as a radical defender of skepticism in the face of the doctrine of a priori certainty, is merely a way of banning doubt from scientific enquiry. Peirce reaffirms the legitimacy of permanent doubt in the process of scientific enquiry – doubt in the sense of indispensable and vital stimulus of thought.

Is Peirce, then, rehabilitating Hume against Descartes? Not at all: Descartes' doubt manifests itself in the question of whether the objects really exist. His doubt was nullified once and for all in the self–certainty of the cogito, as well as by the recognition of ideas as objects of the cogito. Hume's doubt is expressed in the question of whether the objects adhere to the principles of thought, and whether such principles exist at all. Peircean doubt, however, may be construed as the question of whether the objects exhaust themselves in their phenomenology, or if they are not rather signs within a universal system of signs, the deciphering of which is inseparably linked to the participants, themselves sign–producing within this sign system, otherwise known as reality: doubt as an incentive for sign-interpreters, in turn functioning as signs, to explore the new and undefined. The example Peirce uses to explain the function and value of the rehabilitated principle of doubt enables a precise distinction between the positions of Peirce and Hume. As a popular object to demonstrate the power of thought as opposed to the senses, Peirce chooses a melody to exemplify that the anticipatory mental process of thought, which also creates continuity, is necessary to translate the rhapsody of auditory sensations into a melody. We don't "hear" a melody. We hear sounds. We "make" a melody. "Thought is a thread of melody running through the succession of our sensations." (see Peirce, 1985, p. 52).

However, this defense for the authentic role of thought in the development of knowledge is not the point of the "ideas" – that would be less than original. The point is the combination of two objects of thought which may be distinguished as follows:

The first object is the production of a conviction which Peirce calls "belief": "The production of belief is the sole function of thought." (see Peirce, 1985, p. 58). This object in itself does not set the Peircean philosophy of thought apart from Hume's, not even in terms of terminology.

The second object is to make the induced belief act as a "rule for action" which again enables further doubt just when the belief appears to become a dead end. Therefore, this very formal description of the function of thought, which does no more than question the results of the rehabilitation of the right to doubt, represents a remarkable departure from Hume's position.

For Hume, the achieved belief constitutes the "final stopping place" of the thought process, and at the same time the point where doubt is dismissed. Peirce however, wants doubt to be permanent. He provokes thought to continue with its "actions" for the sake of ever innovative progress of scientific enquiry, which he regards as a series of actions triggered by the thought process. By contrast, Hume's scepticism not only strikes towards its own nullification, it also brings scientific enquiry to a standstill. The innovative interest and endless finalism Peirce posits sets him apart from the inductionist tradition in the sense of Hume as well as from deductionist thinking which he critiqued for not venturing beyond the analysis of logical conclusions derived from preconceived axioms. In short: perceived sounds induce listeners to connect them, using their memories or imagination, to create a melody, to organize a symphony ("of thought") based on that melody, and from that symphony perhaps a cosmic synthesis of the arts? With holistic curiosity, the doubt thus defended raises the question of whether the perceived phenomenon is part of a not yet discovered whole.

A look at the methodic principle of "abduction" as introduced by Peirce may help. The inductionist in Hume's tradition compares similar cases, arriving at the conclusion of analogy, which functions much in the same way as the principle of causality does in the Kantian tradition of transcendental deductionism, without, however, as many preconditions. The inductionist concludes from comparison that the ground gets wet whenever it rains. The inductionist then cannot venture further than the "whenever" of conditional phrases and still remain true to his principles. The (transcendental) deductionist can only verify or falsify conditional phrases of this kind through induction. However, he does not deem them reliable enough to serve as the basis of theorems, and for this reason, he questions their value in establishing scientific certainty. He suggests transforming conditional sentences into causal ones (the street gets wet *because* it rains), although under the premise that the cases compared can generally only be interpreted as analogies "because" we think in terms of causality. The transcendental deductionists' burden of proof lies in the difficult question of whether we (always) think in causal terms, i.e., whether "because" sentences can be deduced from an a priori principle of causality.

The abductionist following Peirce's approach goes beyond either of the two disputants. Taking a crucial step which would try the tolerance of the inductionist and, even more so, of the deductionist, he introduces finalism into the principle of causality by interpreting observed phenomena not only as the effects of something, as does the causalist, but regards these same effects as analogous to actions as intended effects or ends. This mode of interpretation is feasible only if the fundamental difference between mind and matter is nullified – matter is "weared out spirit" (see Peirce, 1988, p. 152) – and under the additional condition that mental actions always resemble active processes and actions always pursue effects as their ends. Abduction is the method which positions the phenomena to be interpreted within an open and endless realm of cause and effect – with effects regarded as "signs" for and as objects of ever new hypotheses for the undiscovered causes.

With this, Peirce's position is as far removed from Leibniz's deductionism and the transcendental logic of Kant's deductive system as it is from the inductionism of Hume and Comte. Although one might say that Peirce takes Kant very literally with regard to thought and its being the action of understanding, he demands a consequence from Kantians which Kant himself did not take in the field of scientific methodology, and that is to accept the thesis that thought is not an end in itself, but a "means" to the end of affirming the universal principle of "synechism", which Peirce, in the following text, again uses in both a metaphysical and programmatic sense:

> "When we come to study the great principle of continuity and see how all is fluid and every point directly partakes the being of every other, it will appear that individualism and falsity are one and the same. Meantime, we know that man is not whole as long as he is single, that he is essentially a possible member of society. Especially, one man's experience is nothing, if it stands alone. If he sees what others cannot, we call it hallucination. It is not "my" experience, but "our" experience that has to be thought of; and this "us" has indefinite possibilities." (see Peirce, 1985, p. 64)

This passage is remarkable for the immediacy with which Peirce supplants the principle of "synechism" from the field of natural philosophy to the social sphere. Pragmatic finalism aims at founding a scientific community based on the principle of continuity (see Peirce, 1985, p. 50), thus establishing itself as a paradigmatic community: "Peirce extrapolates," as Habermas notes, "from the experience of the advancement of knowledge to a collective and focused learning proceed of a humankind which has become methodical on the level of organized scientific enquiry." (see Habermas, 1968, p. 119)

The thus formulated ethic of a universal scientific collective as a social avant–garde also has its imperative, which, in accordance with the interpretation of thought as action, addresses reason itself: "Consider what effects, that might conceivably have practical bearings, we conceive the object to our conception to have. Then, our conception of these effects is the whole of our conception of the object." (see Peirce, 1985, p. 62). The imperative becomes clearer within the context of an analogous, more precisely – although clumsily – worded passage from the *Lectures on Pragmatism*:

> "Pragmatism is the principle that every theoretical judgment expressible in a sentence in the indicative mood is a confused form of thought whose only meaning, if it has any, lies in its tendency to enforce a corresponding practical maxim expressible having its apodosis in the imperative mood." (see Peirce, 1935, p. 18)

We can actually conduct the experiment suggested above, i.e., transform a theoretical statement into a conditional sentence with an imperative secondary clause. For instance, we can say that *if* our clear and distinct idea of a thing is necessarily our idea of its (virtual) effects, then the imperative cited above is valid: *Think in a manner in which you regard the object you have named as the effect of a universal causality for which it is a sign.* The thinking individual fulfills the imperative's call for action by co–*operating* in the advancement of science. Hans Apel notes that Peirce wanted to make philosophy operate much in the same way as the natural sciences. This intent of Peirce's makes it easier to understand the example he used to clarify his definition of the concept as *essence of all observable effects of an object*. He reminds us, and rightly so, that the classical concept of "force" is identical to the sum of all of its measurable effects. What interests us here, however, is not the recapitulation of an opinion held by both Newton and Leibniz, but the polemic which, by his reminder, is directed at unnamed supporters of a metaphysic force who make an ontological distinction between the "thing in itself" and its effects, thus isolating them in order to preserve their mystery, a process which led to the distinction between hidden and phenomenal reality. In some respects, this polemic targets Leibniz's doctrine of the "vis activa primitiva" as the dynamic origin of material phenomena (see Leibniz, 1982, p. 6), as well as Kant's concept of the interaction of two "fundamental forces" which are the supposed origins of the spatial forces of attraction and repulsion (see Kant, AA IV, p. 497). Lastly, this critique is aimed at Schelling's early speculations on nature in which he claims that the biological phenomenon of life is grounded in a monistic "formative force" (Bildungskraft) that eludes scientific investigation (see Rudolph, 1993, p. 107). However, these polemics did not estrange Peirce from the more conventional teachings. On the contrary, Peirce always defended Kant, and he even wrote favourably of Schelling in a letter to William James: "My opinions were probably influenced by Schelling, – by all of the phases in Schelling's development, especially his philosophy of nature. I think Schelling is brilliant... so I wouldn't mind it if you called my philosophy a kind of Schellingism transformed in the light of modern physics." (see Peirce, 1985, p. 28)

Peirce dismisses the dualisms he suspects in the said positions – some of which do contain them – in defense of a concept of reality which renders superfluous the distinction between the physical and metaphysical or the intelligible and unintelligible. The message sent in the text's last and perhaps most impressive passage may be summarized as follows: "...real is that whose characters are independent of what anybody may think them to be"(see Peirce, 1985, p. 80). As well as stressing the distance between Peirce and the conflicting currents of nominalism, this sentence documents his position in the long-term aftermath of the universals dispute. He champions a modified realism in which the concepts we develop for intelligible things are real and not mere names, meaning that they are to be regarded as mental concentrates *of* reality. But instead of simply granting this status to concepts such as matter or causality as it was done in the medieval universals dispute, Pearce postulates that our concepts meet these standards (which rise in the course of the dynamic investigative process) much in the same way as the imperative of scientific logic states above. Peirce's position, then, might be described as a kind of "methodic realism":

"This theory of reality is instantly fatal to the idea of a thing in itself, – a thing existing independent of all relation to the mind's conception of it. Yet it would by no means forbid, but rather encourage us, to regard the appearances of sense as only signs of the realities. Only, the realities which they represent would not be the unknowable cause of sensation, but *noumena*, or intelligible conceptions which are the last products of the mental action which is set in motion by sensation." (see Peirce, 1967 [1], p. 261)

The peculiar language Peirce uses here exemplifies the paradoxical nature of his realism with graphic clarity: Reality as it appears to us in the sign of language of phenomena is noumenal; noumena, however, are "products of our mental activity". thus, phenomena themselves are signs of our mental activity: all is mind, while matter is (merely) its (inanimate) expression.

This reverting of Peirce's from methodical to realism to post–metaphysical spiritualism may represent the obstacle for philosophers of science coming to terms with his teachings. In the light of this statement, the imperative of the continuous investigation of effects can be read as an imperative of the constant recognition *of* thought. And indeed, the only thing that keeps Peirce from being confused with Hegel's metaphysics of thought is the fact that he does not employ a dialectic figure to render the relation between phenomenon and noumen *via negationis* (see Peirce, 1967, p. 423). But in the end there is a remarkable resemblance to the nature philosophy of the early and somewhat later Schelling, who also avoided dialectic in his arguments. For Schelling, absolute thought recognizes itself in our thought by "interpreting" nature as an effect of thought.

If Kant's a priorism had not by now become an unacceptable methodological burden on the scientific standards in force today regarding current theories of perception, which claim that the concepts of things can always be created anew instead of things being created by concepts, – if this were not so, defending Kant against Peirce would still be in order for one crucial reason: Kant qualified the perception of things and the inviolable "thing in itself" as the unintelligible source of phenomena, by introducing an incertitude which enables us to grasp nature not only as something we can conceptually determine, but also as a realm of perpetual surprises. In Kant, there is an equal balance between the determinate causality of nature and indeterminate causes of phenomena – the price for this foundation of course, being an intractable dualism that is bound to exasperate any expansive holist. In defense of his claim, however, Kant pointed out that the all–encompassing nature of phenomena must always be assumed to exceed the scope of our understanding. Kant trusted in the ability of the human mind, and its power of reasoning to translate the "given Diversity" into an uniform perception of nature if objects were to be construed with any certainty. However, he did not assume that the world of indeterminate objects in itself is in any way identical with, or much less appertains to, the world of thought.

2   Symbolism or Realism: Is There an Alternative?

In contrast to Cassirer, Peirce's concept of the symbol is an element of a set of diverse types of signs building up a systematical context. A comparison with other competing types of signs helps to clarify the specific difference between the concepts of the symbol in Cassirer and Peirce.

> "A symbol is a representer (*representamen*) that fulfils its function independently from any similarity or analogy to its object and likewise independently from any factual connection with it, but only and uniquely because it would be interpreted as a representer. Every word, sentence and book is an example of this." (see Peirce, 1988, p. 435)

According to the quotation above, it is the independence from the similarity between the sign and the object that distinguishes the symbol from the icon. What distinguishes the symbol from the index is its independence from any actual existence, the total absence of any direct relationship between signs and objects (e.g., in the form of an upward glance of a man, which functions as an index for a balloon that is flying over the city, on the one hand, and that which is looked at, on the other). The symbol presents its object in a very specific way; it is the sign that shows its object – though not insofar as it copies or refers to it, but insofar as it represents it. Of the three types of signs, symbol, icon and index, the symbol has a special ability: it can progressively efface the difference between the sign and the object – though of course only through the interpretative achievement of the interpreter, which in turn can only be produced through the representative capacity of the symbol. It is thus a question of a triadic dynamic interdependence:

> "A symbol differs from the other two types of signs insofar as it only re–presents its object through the interpreter that determines it." (see Peirce, 1988, p. 430)

According to the locus classicus of the Peirceian theory of signs, the symbol appears to be the privileged prototype of the sign:

> "I define a Sign as anything which on the one hand is so determined by an Object and on the other hand so determines an idea in a persons' mind, that this latter determination, which I term the Interpretant of the Sign, is thereby mediately determined by that Object. A Sign therefore has a triadic relation to its Object and to its Interpretant." (see Peirce, 1958, 8.343, p. 232.)

A comparison of these two key sentences shows us that they are both describing the same dynamical semiotic interdependence between sign, object and interpreter from their respective perspectives: the one, from the perspective of the act of interpretation; the other, from the perspective of the object. In both cases, it becomes clear that the sign determines the interpreter and does not merely serve as the mediation between the subject and the object, which it itself has created. In a word, the sign is "somehow always already there".

No better confirmation of the realism of Peirce can be given than this reference to the "somehow always already thereness" of the sign, better than this critique of the nominalism and the belief in a reality independent of the subject. For Peirce, it is a question of "semiotic realism", which in a strict emphatic and antinominalist sense must found the real world as a universe of signs, For Peirce the real is something whose properties are independent of every opinion about it. "Thus we may define the real as that whose characters are independent of what anybody may think them to be." (see Peirce, 1985, p. 80). However, this definition has one reservation: it remains to be clarified whether the interdependence between the sign and the interpreter does not, in general, come about through the opinions that the interpreter has about the sign, and by which he makes the sign a sign. If this is so, then we cannot mean by a sign that which we mean by reality. And if this is so, then it would also seem that Peirce is not a scientist, but rather an ideological realist, as reality for him is not the same as the signs that represent it.

Likewise, Pape writes:

> "...in physical reality which is independent of signs, a development realizes itself which is influenced by the sign in the form of an ongoing series of changes which gives the reality of the material objects a symboloid form (sic). In this way, reality is first determined through the process of interpretation of signs as a symbolic reality." (see Pape in: Peirce, 1988, p. 31)

"Determined", but not discovered we should add. According to Pape it would seem to be necessary to distinguish with Peirce between an independent reality and the constant symbolization of this reality. For Pape, the reality of Peirce is not yet "symboloid".

Is this not simply a displacement of the old duality between the Phenomena and the *Ding-an-sich*? And if not, how is the acceptance of the continuity of the symbolization put forward by this more inclusive thesis of "Synechismus", on the one hand, compatible with the disjunction of the sign and reality, on the other? Declarations of realism and a continued dependence on an unacknowledged nominalism stand in opposition to each other. The possible solution of this tension depends on the symbol's ability to represent. Peirce gives as examples of this ability the "word", "sentence" and "book". Let us consider the "sentence", and for that matter a very specific type of "sentence", which will permit us to compare what has been said with another theorist of the symbol who has attempted to establish a position beyond the conflict of realism and nominalism: namely, the law, or to be more precise, the law of nature. How then does Cassirer interpret the function of the sign of the natural law? Cassirer provides the answer indirectly by the way of a reconstruction of the development from classical to modern physics. For Cassirer, the turn to a "theory of symbols" has already realized itself in the classical paradigm. This transformation is analogous to the movement from substance to function. Cassirer describes this turn, this movement from a "copy theory" to a pure "theory of symbols" in the following way.

According to Cassirer, Galileo and the "classical theory of nature" confirmed the dualism between the objective necessity and sensual quality for the whole of classical thought in physics. Evidently Cassirer indicates that what the "classical" position acknowledges as "objective necessity" is an appropriate image, a copy of reality independent from the subject. He demonstrates the change in the situation through an analysis of Heinrich Hertz's understanding of the sign: if the law has previously been a "copy", it has now become a "symbol". The space between the subjective phenomenon and the objective reality is "redrawn". Neither subjective sensation nor the objectively can any longer claim to coincide with the being of things. In a word: modern physics still grounded in a classical understanding, has moved more and more to a nominalist position. However, to our surprise, modern physics by no means gives up the claim to reality of classical concepts but only "redefined" this claim: this is expressed in the fact that the symbol has been set up as the "border line between the empirical and theoretical". To what extent can we understand Cassirer's concept of the symbol in terms of realism? At the same time, how nominalistically can it be interpreted?

Cassirer seems to attempted to give a definitive answer to this question. It is given in a number of different contexts: in *Substance and Function* in 1910, then more tersely in the third volume of the *Philosophy of Symbolic Forms* in 1929 and in his explicit reference to *Substance and Function* in the 1937 texts in *Modern Physics*, and finally in the polemical treatise of 1938, *The Logic of the Concept of the Symbol*. All these texts are based upon one and the same argument. They all confirm and support one another. The context of the relevant passages in the third volume of the *Philosophy of the Symbolic Forms* seems to be especially helpful in our context. Cassirer refers to the general theory of relativity as representative of the trend in the development of physics according to which the physical world is characterized as a "pure ordered structure" (*Ordnungsgefüge*) of symbols (see Cassirer, 1975, p. 558). In this context, Cassirer portrays the history of physics as the history of increasing physical relativism that culminates for the time being in the history of relativity. It is to be shown, according to him,

> "that certain determinations, which we attached to the object as its conditions, are only definable if we add a certain index to them and thus indicate from what frame of reference they should, as values, be thought. 'Motion' and 'force', 'mass' and 'energy', 'length' and 'duration', are no more something 'in themselves', but they only signify something: and in general, they have a different signification relative to each observer." (see Cassirer, 1975, p. 559)

Cassirer, who seems to interpret this status of the scientific evolution as a manifestation of the successful overthrowing of the classical copy theory by the new theory of symbols, nevertheless still asks whether it would be desirable to have a world concept free from all particularity, in which everything would be related to everything in a single system of relations – a world concept that was beyond all determination and thus represented a perspective "from the point of view of no one"(see Cassirer, 1975, p. 560).

Amazingly enough, Cassirer affirms these questions joining the first and second philosophy – of course not by pleading for the anticipation of a scientifically achievable absolute point of view, but as a hypothetical vanishing point of a process that progresses towards more and more general symbols. In the course of his historical work, Cassirer does not continue to hold the expectation of a process of increasing abstraction of symbolization in general. But as far as one can see he never gives up this idea of increasing abstraction as the price of scientific progress. In Cassirer's later work, the emphasis on the increasing relativism as the consequent way to overcome the concept of substantialization continues to prevail, whereas the idea of progress in sciences as progress in culture continued to recede more and more into the background, at least until 1933.

The result is particularly interesting for the determination of the concept of symbols: "force", "energy" or "mass" are bearers of meaning. Through them the physicist interprets a phenomenon, which is at the same time represented by them. Symbols are the result of the transfer of meaning. The more appropriately they represent the meaning, the more independent they are from the object to which they refer and from the subject who marks the object with them. Symbolism recommends itself as the alternative to objectivism, subjectivism as well as substantialism. Cassirer's remark seems to mean that symbols would have their place at the border between the empirical and theoretical.

With Peirce, Cassirer joins, in a certain phase of historical work, the appraisal of the history of culture as the process of increasing complexity of symbolization. Nevertheless, the difference between the two positions remains striking. This difference manifests itself in their different views of the productive power of the symbolic concept:

> "The cultural development of man transforms... the part of reality
> that is accessible to him and himself into elements in a large process
> of signs that controls itself." (see Pape in: Peirce, 1988, p. 31)

Such a difference between the isolated segments and the rest of reality is as foreign to Cassirer as the identification of symbolism and realism (as it has been explained above). For Cassirer, the segment and reality fall together. On this point he remains the radical heir of Kant's philosophy of finitude, with the significant difference that he does not need (nor accept) a remaining dualism between a reality "for us" and a reality "in itself". Peirce, on the contrary, tends (and he will pay the price for his departure from the unsatisfactory conditions of the subjective nominalism) toward a semiotic realism, which would have a difficult time defending itself against the reproach that it represents a variant of metaphysical substantialism.

References

Cassirer, Ernst (1975): *Philosophie der symbolischen Formen*, Vol. 3 (Darmstadt, Wissenschaftliche Buchgesellschaft)

Habermas, Jürgen (1968): *Erkenntnis und Interesse* (Frankfurt, Suhrkamp)

Kant, Immanuel: *Metaphysische Anfangsgründe der Naturwissenschaft*, AA IV

Kant, Immanuel: *"Das Jahr 69 gab mir grosses Licht"*, AA XVIII

Leibniz, Gottfried Wilhelm (1982): *Specimen Dynamicum*, ed. by Dosch, H.G./Most, G.W./Rudolph, E.(Hamburg, Felix Meiner)

Oehler, Klaus (1993): *Charles Sanders Peirce* (Munich, Beck)

Peirce, Charles S. (1935): *Collected Papers, Vol. V*, ed. by Hartshorn, Charles and Weiss, Paul (Cambridge (Mass.)/London, Belknap Press)

Peirce, Charles S. (1958): *Collected Papers, Vol. VIII*, ed. by Burkes, Arthur W.(Cambridge, Harvard University Press)

Peirce, Charles S. (1967): *Schriften II. Zur Entstehung des Pragmatismus*, ed. by Apel, Karl Otto (Frankfurt, Suhrkamp)

Peirce, Charles S. (1985): *How to make our ideas clear / Ueber die Klarheit der Gedanken*, Introduction, transl. and comm. by Oehler, Klaus (Frankfurt, Klostermann)

Peirce, Charles S. (1986): *Schriften I. Zur Entstehung des Pragmatismus*, ed. and transl. by Klösel, Christian and Pape, Helmut (Frankfurt, Suhrkamp)

Peirce, Charles S. (1988) *Naturordnung und Zeichenprozess. Schriften über Semiotik und Naturphilosophie*, ed. and introd. by Pape, Helmut (Aachen, Alano)

Rudolph, Enno (1989): Entelechie. Zeit und Individuum bei Leibniz. In: von Weizsäcker, Carl-Friedrich/Rudolph, Enno (Eds.) (1989):*Zeit und Logik bei Leibniz. Studien zu Problemen der Naturphilosophie, Mathematik, Logik und Metaphysik*, Stuttgart, Klett-Cotta)

Rudolph, Enno (1993): Die Natur als Subjekt. Zur Leibnizrezeption des frühen Schelling. In: Gloy, Karen/Burger, Paul (Eds.): *Die Naturphilosophie im Deutschen Idealismus* (Stuttgart-Bad Cannstatt, Frommann-Holzboog)

5. Heinrich Hertz and the Concept of a Symbol

Andreas Hüttemann

In a recently published article A. Nordmann highlighted the fact that Hertz considered it as the greatest pleasure of scientific research to be "alone with nature" and to learn "directly from nature" (see Nordmann, 1998, p. 156). Hertz contrasts this being on his own with nature with the "disputes about human opinions views and demands". (see Nordmann, 1998, p. 156). It is this contrast between nature on the one hand and human beliefs etc. on the other that is fundamental for his central epistemological pursuit: the attempt to sort out which features of our theories can be attributed to nature as opposed to those which depend on us. I will discuss the various writings in which Hertz touches this subject.

1   Hertz on Helmholtz's Theory of Signs

The context in which Hertz discusses his epistemological question for the first time is Helmholtz's theory of signs. Hertz was particularly impressed by the claim that the structure of the eye partly determines what is perceived.

In a newspaper article on Helmholtz's 70th birthday in 1891, Hertz calls Helmholtz's research in physiology one of his main achievements. He characterizes this research in terms of questions that eventually lead to his own interests:

> "How is it possible for vibrations of the ether to be transformed by means of our eyes into purely mental processes which apparently can have nothing in common with the former; and whose relations nevertheless reflect with the greatest accuracy the relations of external things? In the formation of mental conceptions what part is played by the eye itself, by the form of the images which it produces, by the nature of its colour-sensations, accomodation, motion of the eyes, by the fact that we possess two eyes? Is the manifold of these relations sufficient to portray all conceivable manifolds of the external world, to justify all manifolds of the internal world?"[1]

These questions concerning visual perception can be asked with respect to all knowledge. Thus Hertz continues:

---

[1] See Hertz (1896) p. 336. It should be noted here that Hertz uses the concept of an *image* to characterize those items that are not only determined by nature but also by some features of the eye. This is noteworthy because Helmholtz himself prefers the notion of *sign* in this context. Helmholtz (1896), p. 586.

"We see how closely these investigations are connected with the possibility and legitimacy of all natural knowledge. The heavens and the earth doubtless exist apart from ourselves, but for us they only exist in so far as we perceive them. Part of what we perceive therefore apertains to ourselves: part only has its origin in the properties of the heavens and the earth. How are we to separate the two?" (See Hertz 1896, p. 336/7.)

Nature is accessible to us through perception only. The question concerning the exact borderline between what among our representations, perceptions or ideas is grounded in nature in contrast to what is determined by ourselves is the epistemological question that Hertz deals with in his writings on both electrodynamics and mechanics – his central epistemological question. Even though this newspaper article was written in 1891, considering the fact that he connected his question with the theory of signs, which he presumably came across much earlier, combined with his remarks quoted at the very outset of the paper, we might very well conclude that this epistemological question was something that was on his mind for a long time before the 1890s.

In what follows I intend to show that Hertz's considerations concerning the comparison of theories and the introduction of the concepts of symbol and image ought to be seen as attempts to answer this central epistemological question. Hertz tried to determine the borderline between what we can legitimately attribute to nature and what has to be counted as our own construction.

Hertz's reflections on the comparison of electrodynamic theories are a first attempt to invent an appropriate terminology to solve this question. It does not yet rely on the concepts of symbol and image.

2   The Comparison of Electrodynamic Theories

In the introduction to *Electric Waves*, Hertz indicates what kind of understanding he has gained at the beginning of the 1890s of his own experimental and theoretical research. On the basis of the work done by Helmholtz, Hertz had tried to compare the theories of Weber, Helmholtz and Maxwell. His experimental research in this area was decisive for the ultimate acceptance of Maxwell's theory of electrodynamics.

The second part of Hertz's introduction summarizes what he takes to be his main achievements with regard to his theoretical investigations of electric waves. His starting point is the question "What is it, that we call the Faraday-Maxwell theory?" (see Hertz, 1962, p. 20). In order to answer his question he draws a distinction between the representation (Darstellung) and the content (the English text uses "inner significance" as a translation for "Inhalt") of a theory. Hertz distinguishes three representations of Maxwell's theory: Maxwell's representation, the representation as a limiting case of

Helmholtz's electrodynamics and his own. All of these are representations of the *same* content. What all of these representations have in common is the system of Maxwell's equations. For a representation to be a representation of Maxwell's theory, it is a both necessary and sufficient condition to yield these equations.[2] This is why he famously answered to the question "What is Maxwell's theory?" as follows: "I know of no shorter or more definite answer than the following: – Maxwell's theory is Maxwell's system of equations." (Hertz, 1962, p. 21). That, however, does not yet answer the question as to the nature of a representation of a theory. In the particular case at hand, Hertz contrasts the mathematical relations with the *physical significance* of Maxwell's claims (Hertz, 1962 p. 20). The representations thus add physical significance to the system of equations. They do this by invoking physical conceptions (Vorstellungen) such as "pictures of electrified atoms" or "concrete representations (Vorstellungen) of the various conceptions as to the nature of electric polarisation, the electric current etc." (Hertz, 1962, p. 28). Elsewhere we read:

> "Maxwell originally developed his theory with the aid of very definite and special conceptions as to the nature of electrical phenomena. He assumed that the pores of the ether and of all bodies were filled with an attenuated fluid, which, however, could not exert forces at a distance." (Hertz, 1962, p. 27).

A representation of a theory adds physical significance to abstract concepts such as polarization or electricity by corellating them, for instance, to more familiar concepts or pictures of other branches of physics.

It is important not to misconstrue Hertz's notion of a representation and its conceptions. Hertz was not a precursor of the so-called received view of theories, as is sometimes claimed.[3] According to the received view a physical theory comprises two essential features, an abstract uninterpreted mathematical calculus (e.g., Maxwell's equations) and a set of correspondence rules that

---

[2] "Every [representation of a] theory which leads to the same system of equations, [...] I would consider as being a form or special case of Maxwell's theory; every theory which leads to different equations, [...] is a different theory" (Hertz, 1962, p. 21). Instead of "theory" Hertz should have used "representation of a theory" at the beginning of this passage. Hertz does not always live up to the criteria that he requires theories to have (see Sect 6) while presenting his views on their nature. For instance, he uses not only "theory" in places where he clearly means "representation of a theory"; instead of this latter expression there are a lot of further expressions that he apparently regards as synonymous, such as "form", "special case" (in the above quotation), and "Fassungen" which has been translated as "modes of representation" (p. 21). There is also the notion of "standpoint", whose relation to the above is not entirely clear. (There are *four* standpoints but only *three* representations with respect to Maxwell's system of equations).

[3] This view is held by de Agostino (1998), p. 89–102, see especially p. 90/91. It has been criticized by Heidelberger (1998), p. 9–24, especially p. 19/20.

serves to link the theoretical terms to experimental procedures.[4] The correspondence rules provide a (partial) interpretation of the theoretical terms; they confer empirical significance. Maxwell's equations are taken to lead to predictions of phenomena independent of any representation as the following passage indicates:

> "Every theory which leads to the same system of equations, *and therefore comprises the same possible phenomena*, I would consider as being a form or special case of Maxwell's theory; every theory which leads to different equations, *and therefore to different possible phenomena*, is a different theory." [my emphasis] (Hertz, 1962, p. 21).

The upshot is that we should not consider a representation of a theory as an interpretation in the sense of the received view of theories.

Having said that, we are still left with the question of why we need representations of theories and the images, models, etc., they come along with. Ultimately Hertz does not provied a clear-cut answer to this question. He is clearly suspicious of models, images, etc. In fact, Hertz defines his own objective in his theoretical papers, as he outlined in the introduction to *Electric Waves*, as the attempt to develop a representation of the system of Maxwell's equation that can do without pictorial conceptions (Vorstellungen) as far as possible:

> "I have [...] endeavoured in the exposition to limit as far as possible the number of those conceptions which are arbitrarily introduced by us, and only to admit such elements as cannot be removed or altered without at the same time altering possible experimental results." (Hertz, 1962, p. 28).

Hertz claims that he has removed all pictures etc. that he could possibly remove. Thus, it is apparently impossible to eliminate all of these "elements". Hertz does not give us a reason why it is impossible to do entirely without images, models, etc., nor does he provide a conjecture as to their positive use (Hertz, 1962, p. 28). In *Principles of Mechanics* he links this indispensability to the structure of the human mind.

Be that as it may, Hertz's attempt to do without pictorial conceptions relies on a distinction he draws. On the one side are those features which we introduce arbitrarily into a theory; on the other side are features whose modification yields a modification of possible experience. This distinction coincides with the distinction that characterizes his main epistemological question as becomes apparent in the following passage:

> "It is true that in consequence of these endeavours, the theory acquires a very abstract and colourless appearance. [...] But scientific accuracy requires of us that we should in no wise confuse the simple

---

[4] For a detailed discussion and a presentation of the development of the received view of theories see Suppe (1977), p. 1–241.

and homely figure, as it is presented to us by nature, with the gay garment which we use to clothe it. Of our own free will we can make no change whatever in the form of the one, but the cut and colour of the other we can choose as we please." (Hertz, 1962, p. 28).

Hertz distinguishes two factors that determine a theory, nature herself and us. What nature contributes turns out to be mutable only at the cost of a change in the description of possible phenomena, whereas what *we* contribute is *arbitrary* and by implication does not yield a change in the phenomena if modified.

Hertz conceives his own papers in theoretical electrodynamics as attempts to sort out these two features. Thus, he attempts to give an answer to his main epistemological question. It is not only his explicitly epistemological remarks that deal with this question but also his theoretical work in physics – at least according to his self-assessment.

3   The Objective of *Principles of Mechanics*

In a paper that Hertz delivered in 1889, he refers to the question concerning "the essence, the properties of the space-filling medium – the ether, his structure, his rest or movement, his infinity or limitedness" as the question of supreme importance in physics. "The question whether everything that is, has been created out of ether isn't any longer out of the reach of today's physics. These things are the ultimate aim of our science, physics." (Hertz, 1894, p. 354). To achieve the aim of physics, that is, to explain the essence of the ether, the equations of motions have to be reduced to the laws of mechanics. This reduction, however, cannot be successful, as Hertz remarks in the preface to *Principles of Mechanics* "until we have obtained a perfect agreement as to what is understood by this name [of laws of mechanics]". (Hertz, 1956). Thus the elucidation and explanation of the foundations of mechanics are a necessary prerequisite for the realization of the ultimate goal of physics, that is, the investigation of the ether. His research in mechanics concerns solely this prerequisite, not, however, the physics of the ether itself, as he points out to his former Strasbourg colleague E. Cohn in 1891:

> "What you have been hearing about my work by way of Halle is unfortunately without any basis and I don't know how this opinion originated. I haven't worked on the mechanics of the electrical field at all, and haven't anything about the motion of the ether. This past summer I reflected a lot about ordinary mechanics, but I don't remember speaking about this in Halle at all." (Nordmann, 1998, p. 160).

The question arises as to what kind of problems in "ordinary" mechanics Hertz intended to solve. As the letter to Cohn continues, it becomes obvious that he is again attempting to draw an epistemological borderline.

"Here I would like to put some things in order and to determine
the order of concepts in such a manner that one can see more clearly
what is definition and what is empirical fact, e.g., in the concept of
force, of inertia, etc." (Nordmann 1998, p. 160).

Hertz's attempt to distinguish definition and empirical fact is again aimed at
answering his central epistemological question – now with respect to mechanics. It is *Principles of Mechanics* as a whole and not just the introduction
that has to be taken to be an elaborated attempt to draw the distinction
between what can be attributed to nature herself and what not.

4   The Concept of a Symbol or Image
in *Principles of Mechanics*

The concept of a symbol is used only once in *Principles of Mechanics* but
Hertz makes it clear he considers it to be synonymous with the concept of an
image that he uses throughout.[5] These concepts are introduced by Hertz to
characterize what we rely on when we make predictions, *i.e.*, what he calls the
"most direct, and in a sense most important, problem which our conscious
knowledge of nature should enable us to solve" (see Hertz 1956, p. 1):

"We form for ourselves images or symbols of external objects; and
the form which we give them is such that the necessary consequents
of the images in thought are always the images of the necessary consequents in nature of the things pictured." (Hertz, 1956, p. 1).

The images (Bilder) Hertz speaks of are also called "conceptions" (Vorstellungen) of things:

"The images we here speak of are our conceptions of things. With
the things themselves they are in conformity in one important respect,
namely in satisfying the above-mentioned requirement. For our purposes it is not necessary that they should be in conformity with the
things in any other respect." (Hertz. 1956, p. 1/2).

The first thing to be noted is that Hertz makes use of the concept of an
image in a narrow and in a broad sense. Images in the narrow sense are
parts of theories that refer to particular things in nature. This is the sense
in which the concept of a symbol or image is used in the above quotation.
When he compares the different images of ordinary mechanics, it is rather
theories as a whole that he has in mind. The above-quoted requirement for
symbols or images is valid both for the narrow sense as well as for the broad,
as becomes clear directly after the introduction, where he exclusively deals
with the broad sense of an image.

---

[5] There are more occurrences in the english edition. Thus on p. 139 the translator
uses "symbol" for the german "Zeichen".

What an image ought to aim at is therefore clearly determined: the prediction of future (and presumably the explanation or retrodiction of past) events. What Hertz refers to as the "conformity" of the consequents of the images of things with the consequents of the things pictured would nowadays be referred to as the empirical adequacy of theories.

How do we compare the consequents of images with the consequents of things? Let us start with the constitutive elements of images. Hertz refers to *fundamental ideas* and *principles*, connecting them as the main elements that are characteristic for a particular image. Principles of mechanics are defined as

"[a]ny selection from amongst such and similar propositions, which satisfies the requirement that the whole of mechanics can be developed from it by purely deductive reasoning without any further appeal to experience." (Hertz, 1956, p. 4).

Thus the whole experiential input of a theory has to be captured by its principles. Hertz's own principle is therefore not discussed in the first part of his book, which does not concern itself with experience at all, but in the second.

The examples of images Hertz discusses in *Principles of Mechanics* are the customary representation of mechanics which is characterized through the fundamental ideas of space, time, mass and force as well as Newton's laws of mechanics and D'Alembert's principle. The ideas of space, time, mass and energy together with Hamilton's principle constitute the "energetical" image. Hertz's own image presupposes just three fundamental ideas, space time and mass – plus a fundamental law that serves as his principle.

Thus far we have dealt with principles and fundamental ideas of images. We may also deduce propositions from the fundamental ideas and the principles, i.e., the consequents of our images. The question arises as to how we are able to check the conformity of the latter with the consequents of the things pictured. The first book of *Principles of Mechanics* does not deal with this problem; it treats the fundamental ideas and introduces definitions without making any reference to nature. "The subject matter of the first book is completely independent of experience." (Hertz, 1956, p. 45). It is only in the second book that such a connection is established. At the beginning of the second book Hertz introduces three *rules* (Festsetzungen) for his fundamental ideas. The first of these rules concerns time:

"Rule 1. We determine the duration of time by means of a chronometer, from the number of beats of its pendulum. The unit of duration is settled by arbitrary convention." (Hertz, 1956 p. 140).

There are similar rules for space and mass. Even though these rules remind one of correspondence rules that give meaning to the concepts in question Hertz does not conceive them as such. Rather, he thinks of them as providing definite and determinate values for a determinable. If Hertz had – anachronistically – conceived of the rules as correspondence rules, they would have served to determine the meaning of, say, time. That is something Hertz does not consider to be lacking. The conceptual structure, the fundamental ideas have been outlined in the first book – they do not need an interpretation. Hertz himself puts it like this:

> "The three foregoing rules are not new definitions of the quantities time, space and mass, which have been completely defined previously. They represent rather the laws of transformation by which we translate external experience, i.e., concrete sensations and perceptions into the symbolic language (Zeichensprache) of the images of them which we form [...] and by which conversely the necessary consequents of this image are again referred to the domain of possible sensible experiences." (Hertz, 1956 p. 141).

What is important is that it is not only the fundamental ideas and the principles but also these rules that are constitutive for the concept of an image (or symbol):

> "Thus only through these rules can the symbols (Zeichen) time, space and mass become parts of our images of external objects. Again, only by these three rules are they subjected to further demands than are necessiated by our thought." (Hertz, 1956 p. 141).

Thus, it is only with the help of these rules that the images become images of external things. It is not that Hertz thinks that his fundamental images lack all meaning in the absence of rules. They lack empirical significance. Not being an empiricist, he does not equate these two.

5 Images and Models

At one point in the introduction to *Electric Waves* Hertz explicates the concept of an image (or symbol) by referring to models. This is interesting because Hertz gives an explicit definition of a model in the second part of *Principles of Mechanics*. This definition of a model and its relation to the notion of an image highlights the fact that images are underdetermined by the aim or criterion of empirical adequacy.

A material system is a model, as defined by Hertz, if it stands in the following relation to another material system (Hertz, 1956, p. 175):

> "**Definition.** A material system is said to be a dynamical model of a second sytem when the connections of the first can be expressed by such coordinates as to satisfy the following conditions:
> (1) That the number of coordinates of the first system is equal to the number of the second.

(2) That with a suitable arrangement of the coordinates for both systems the same equations of condition exist.

(3) That by this arrangement of the coordinates "the expression for a magnitude of a displacement agrees in both systems"."

According to this definition the model relation is symmetric. Two systems are models of each other. Furthermore, material systems are not completely characterized through the modelrelation:

"A system is not completely determined by the fact that it is a model of a given system. An infinite number of systems, quite different physically, can be models of one and the same system. Any given system is a model of an infinite number of totally different systems. For the coordinates of the masses of the two systems which are models of one another can be quite different in number and can be totally different functons of the corresponding coordinates." (Hertz, 1956 p. 176).

It is this concept of a model that Hertz uses to characterize images. At the outset of the introductio (Hertz, 1956, p. 1) he had already pointed to a connection that he explicates in more detail in the second part:

"The relation of a dynamical model to the system of which it is regarded as a model, is precisely the same as the relation of the images which our mind forms of things to the things themselves. For if we regard the condition *(Zustand)* of the model as the representation of the condition of the system, then the consequents of this representation, which according to the laws of this representation must appear, are also the representation of the consequents which must proceed from the original object according to the laws of this original object. The agreement between mind and nature may therefore be likened to the agreement between two systems which are models of one another." (Hertz, 1956, p. 177).

Hertz does not identify image and model, it is rather the *relation* between two material systems that is the same as the *relation* between the system and the image. As Hertz uses these terms, for an image to be a model of a *material* system it has to be a material system not a mental system.

The identity of the relations entails that just as a material system is not completely determined if it stands in a model relation to another system, an image is similarly *underdetermined*. It is underdetermined by the conformity requirement, i.e., by the requirment of empirical adequacy.

6   Criteria for the Evaluation of Images

The requirement of empirical adequacy does not determine an image completely. Besides this criterion, which Hertz sometimes calls "correctness", he introduces two further criteria for the evaluation of images. These are the criteria of *admissibility* and *appropriateness*. An image is admissible if it does not contradict the laws of our thought, i.e., if it is logically consistent. An image can be appropriate in two respects. It can be more appropriate than another image if it is more distinct. This is the case if it "pictures more of the essential relations of the object" than its competitor (Hertz, 1956, p. 2). Also, an image may be more appropriate than another if it is simpler, i.e., if it contains "in addition to the essential characteristics, the smaller number of superfluous or empty relations." (Hertz, 1956, p. 2).

The three criteria of correctness (empirical adequacy), admissibility (logical consistency) and appropriateness that Hertz invokes in order to compare the different images of mechanics are definitely linked to three factors that, according to Hertz, determine an image.

> "What is ascribed to the images for the sake of appropriateness is contained in the notations, definitions, abbreviations, and, in short all that we can arbitrarily add or take away. What enters into the image for the sake of correctness is contained in the results of experience, from which the images are built up. What enters into the images, in order that they may be permissible, is given by the nature of mind." (Hertz, 1956, p. 2/3).

Hertz requires that a presentation (Darlegung) of an image ought to analyze to what extent the image satisfies these criteria. This requirement is tantamount to asking for an answer to his central epistemological question. What are the features of an image that depend on nature (or experience) and what are the ones that depend on us? Hertz distinguishes in *Principles of Mechanics* two kinds of theoretical features that cannot be attributed to nature. First, there are those features for which we are responsible willingly (definitions, abbreviations, etc.) and, second, there are those features for which we are responsible unwillingly, *i.e.*, the conformity of the image to the laws of thinking. It is this additional distinction of arbitrary and non-arbitrary elements of a picture that had been introduced neither in Helmholtz's theory of signs nor in Hertz's earlier conception of scientific theories in the introduction of *Electric Waves*. Thus we have the following links between criteria on the one hand and determining factors of a theory on the other: nature (experience) – correctness; necessities of thought – admissibility; arbitrary choice – appropriateness.

Hertz is critical of the fact that the traditional (received) image of mechanics has never been analyzed with respect to these elements:

> "It still fails to distinguish thoroughly and sharply between the elements in the image which arise from the necessities of thought, from experience and from arbitrary choice." (Hertz, 1956, p. 8).

In *Principles of Mechanics* Hertz attempts to present an image of mechanics such that it becomes apparent what the determining factors of his image are. Given this aim one would expect *Principles of Mechanics* to consist of three books – one for each of the determining features. It does, however, contain only two booksn which reflects the difficulty of sorting out the determining factors (see Sect. 7). The first book is devoted entirely to those elements that are dependent on us. He presents definitions of fundamental concepts or ideas that are arbitrary in the sense that he could have chosen other definitions. He then deduces with the help of our laws of thinking propositions that are not yet connected to anything in nature. It is only in the second book that predictions become possible – thanks to the rules (Festsetzungen) and his fundamental principle (which comprises the whole empirical content of the theory). The latter is the factor that represents to what extent the image depends on nature or experience.

7 Problems

There are some problems that Hertz has to face in carrying out his analysis in order to answer his central epistemological question. One problem is that it does not seem to be altogether easy to isolate certain features of a theory as being dependent on exactly one determining factor. This becomes apparent if one looks at his criticism of the competing images of mechanics. As he points out, these images are not appropriate in the sense of being simple, i.e., they postulate too many (empty) relations. This in turn yields logical inconsistencies:

> "But we have accumulated around the terms 'force' and 'electricity' more relations than can be completely reconciled amongst themselves. We have an obscure feeling of this and want to have things cleared up. Our confused wish finds expression in the confused question as to the nature of force and electricity. But the answer which we want is not really an answer to this question. It is not by finding out more and fresh relations and connections that it can be answered; but by removing the contradictions existing between those already known, and thus perhaps by reducing their number. When these painful contradictions are removed, the question as to the nature of force will not have been answered; but our minds, no longer vexed, will cease to ask illegitimate questions." (Hertz, 1956, p. 7/8).

If we were to remedy this situation by postulating less relations, we would at the same time enhance appropriateness and admissibility. However, if the criteria are connected in this way it seems difficult to attribute certain features of the theory to exactly one of the determining factors (which are definitely connected to exactly one of these criteria).

A second problem is that Hertz is not entirely clear about the relative merits of these criteria. In the introduction to *Principles of Mechanics* he maintained that correctness or empirical adequacy is the most important aim that our knowledge of nature can achieve. What then is the status of the other criteria? Are admissibility and appropriateness merely pragmatic virtues? Maybe correctness presupposes admissibility. Is it the case that theories that are more appropriate are more likely to be correct? Why should a physicist be interested in appropriateness if that were not the case?

Also, Hertz is not quite clear about why his favourite image outdoes the competitors.[6] At one place it seems to be admissability (logical consistency) that is the most important criterion in virtue of which his image surpasses the competitors.

> "This merit of the representation I consider to be of greatest importance, indeed of unique importance." (See Hertz 1956, p. 33.)

However, later on he claims that the traditional images and his own are on a par with respect to admissability and it is rather the correctness that will have to decide between the two pictures:

> "We shall then have as our sole criterion the correctness of the images [...] and here it is important to observe that only one or the other of the two images can be correct: they cannot both at the same time be correct." (Hertz, 1956, p. 40).

This leaves us with the impression that we have two theories of mechanics that differ in their predictions.

8   Conclusion

Conceiving physical theories as images or symbols is a means to answer Hertz's central epistemological question. Theories as images or symbols owe *some* of their features to what they stand for – in this case nature or experience. However, they also owe some of their features to those who produce or construct them. These features are partly non-arbitrary and partly arbitrary. For the further development of the concept of the symbol, Hertz's insistence on the latter factor became important. Thus, Cassirer always referred to the constructive element in Hertz's account of images or symbols.[7]

---

[6] This is a point Alfred Nordmann has highlighted (Nordmann, 1998, p. 161).

[7] Cassirer called symbols in the sense of Hertz "konstruktiver Entwurf" (Cassirer, 1954, p. 25).

References

Cassirer, E. (1954): *Philosophie der symbolischen Formen* part III (Darmstadt), 25.

de Agostino, S. (1998): "Hertz's View on the Method of Physics: Experiment and Theory Reconciled?" in Baird, D., Hughes, R. I. G. and Nordmann A. (Eds.): *Heinrich Hertz: Classical Physicist; Modern Philosopher* (Dordrecht), 89–102.

Heidelberger, M. (1998): "From Helmholtz' Philosophy of Science to Hertz' Picture Theory" in Baird, D., Hughes, R. I. G. and Nordmann A. (Eds.): *Heinrich Hertz: Classical Physicist; Modern Philosopher* (Dordrecht), 9–24.

Hertz, H. (^{2}1894): "Über die Beziehungen zwischen Licht und Elektrizität" in Hertz, H.: *Gesammelte Werke* Band I (Leipzig), 339–354.

Hertz, H. (1896): "Hermann von Helmholtz" in *Miscellaneous Papers* (London), 332–340.

Hertz, H. (1956): *The Principles of Mechanics* (New York).

Hertz, H. (1962): *Electric Waves* (New York).

Nordmann, A. (1998): "'Everything could be different': *The Principles of Mechanics and the Limits of Physics*" in Baird, D., Hughes, R. I. G. and Nordmann A. (Eds.): *Heinrich Hertz: Classical Physicist, Modern Philosopher* (Dordrecht), 155–171.

Suppe, F. (1977): "The Search for Understanding of Scientific Theories" in Suppe, F. (Ed.): *The Structure of Scientific Theories*, Urbana (2nd ed.), 1–232.

Helmholtz von, H.: *Handbuch der physiologischen Optik* (Hamburg/Leipzig) 1896 (2nd ed.).

Part III

On the Symbolic Structure of Physics

6. Shifting Symbolic Structures and Changing Theories: On the Non-Translatability and Empirical Comparability of Incommensurable Theories*

Martin Carrier

1 Symbolic Descriptions and the Choice of Conceptual Structures

Scientific theories in the mature sciences do not consist of a collection of observational regularities, but are intended to capture what lies behind the phenomena. Theoretical concepts transcend the realm of immediately perceptible objects and processes; they are removed from the phenomenal world in that their referents typically remain hidden to the unaided senses. There are good reasons for conducting science by constructing theories rather than collecting observations. The history of science suggests that providing unified explanations of large areas of phenomena requires concepts that are detached from what can directly be experienced. A frequently mentioned example is taken from the history of thermodynamics. Drawing on observational quantities, one arrives at empirical regularities of the behavior of gases. Experience produces a number of lawful relations among quantities such as pressure, volume, and temperature. Adopting, however, a theoretical perspective and positing that the behavior of gases results from molecular collisions, governed by the laws of statistical mechanics, first, establishes connections among these relations, and, second, suggests corrections of them. Theoretical reasoning demonstrates that the empirically established regularities hold only approximatively, and moreover it indicates how the description is to be improved. Theoretical terms promote scientific understanding more effectively than observational concepts. Recourse to unobservables allows for a more thorough and deep-searching account of the phenomena (see Hempel, 1966, Chap. 5).

On the other hand, the relationship between theoretical concepts and the objects or processes in nature is complex and tenuous. Observational concepts such as pressure or volume are robust in this respect; their counterparts can be exhibited in experience. By contrast, the supposed referents of concepts like "molecular collisions" or "atomic interaction" cannot. The terms appropriate to theoretical explanation cannot simply be read off from

---

* I am grateful to Giora Hon, Paul Hoyningen-Huene, and Thomas Kuklinski for valuable advice.

the phenomena. Put more positively, scientists enjoy some freedom of concept formation in constructing theories. Suitable concepts need to be crafted by human ingenuity; they are not imposed by nature.

It follows that a given realm of phenomena may be subject to disparate theoretical approaches. The rather abstract and detached nature of scientific concepts makes room for alternative explanations of the same set of data. This feature in turn creates the possibility that a scientific discipline undergoes a drastic theoretical alteration in which the formerly received view is replaced by a conceptually disparate one. Thomas S. Kuhn famously claimed that science is prone to display such theoretical about-faces or "scientific revolutions". In order for one such comprehensive theoretical tradition or paradigm to supplant another one, the former needs also to account for the data that the latter explains. Since the rivaling theories are supposed to be conceptually incongruous, they are likely to entail divergent explanations of the observations they jointly address.

This feature is important for elucidating the nature of scientific concepts in general. First, such concepts are symbolic in kind. They fail to depict the entities they are intended to represent (in the way a drawing pictures an object), nor do they exhibit any physical relationship to them (in the way in which smoke denotes the presence of fire). Couched in Peircean terms, scientific concepts are neither icons nor indexes: they are not similar to the entities they refer to, nor are they causally related to them (see Chap. 4). Observational and theoretical concepts alike are symbols whose representational powers rather derive from human convention than from natural virtues. Moreover, theoretical concepts are supposed to represent unobservable objects or processes. The given revolutionary scenario entails a situation where divergent symbolic structures compete for acceptance by a scientific community. Consequently, at least one of them, or maybe both, inevitably fails to refer to real entities. The existence of scientific revolutions raises doubts as to the representational virtues of theoretical terms.

Conceptual disparity between theories highlights the problem of reference inherent in the use of symbols in general. In the case where two (or more) explanations of the same phenomena could be given that are conceptually divergent and yet empirically equivalent, the reference of the terms involved would appear doubtful. I refrain from addressing this problem in full and rather restrict myself to the first relevant aspect, namely, the nature and impact of the conceptual divergence involved in scientific revolutions. This problem is subsumed under the heading of incommensurability.

2   Meaning, Theoretical Context, and Adequate Translation

Incommensurability is among the catchwords of the later 20th century philosophy of science. The notion of incommensurability in the non-geometrical sense relevant here was simultaneously introduced by Kuhn and Paul K. Feyerabend in 1962 (see Kuhn, 1962, p. 103; Feyerabend, 1962 p. 58). Kuhn conceived incommensurability to be a contrast between paradigms or comprehensive theoretical traditions that transcends mere incompatibility. The adoption of a new paradigm entails the restructuring, as it were, of the relevant universe of discourse; the adherents of the two paradigms tend to talk past one another. In particular, incommensurability is intended to express that, first, disparate concepts are employed in each of the theories at hand; second, distinct problems are tackled; third, the suggested problem solutions are evaluated according to different standards; and, finally, perceptions are structured differently (see Kuhn, 1962, p. 103–110, 148– 150). Feyerabend, by contrast, focused on the "inexplicability", that is, the non-translatability of a term taken from one theory into the conceptual framework of another one incompatible with the first. Feyerabend's example is the failure to fit the notion of "impetus" into Newtonian mechanics (see Feyerabend, 1962, p. 52–62).

While the initial use of the term "incommensurability" varied significantly, it was subsequently restricted to denote the non-translatability of concepts or statements from different, strongly contrasting theories. In the following I exclusively address this more limited notion of incommensurability that is sometimes called "semantic incommensurability". In his later years, Kuhn attempted to trace semantic incommensurability back to changing assumptions about what is alike or what is of the same kind. Incommensurability was supposed to represent a translation failure resulting from conflicting structures of kinds, scientific, natural, or otherwise (see Kuhn, 1983, p. 680–684; Kuhn, 1990, p. 5; Kuhn, 1993, p. 315–319; see also Carrier, 1994, p. 7–8).

My aim is to give a systematic reconstruction of the nature and impact of semantic incommensurability. I start off from Kuhn's later notion, i.e., non-translatability of concepts due to incompatible structures of kinds, and I give a brief "rational reconstruction" of this notion. Underlying Kuhn's approach to translation is the "theoretical context account of meaning" or "semantic holism". My reconstruction of incommensurability proceeds from the context account as the central premise. In contrast to Kuhn's own presentation of the subject matter, I do not address shifting structures of scientific kinds directly, but only derivatively as resulting from the substantial alteration of scientific laws. A law or generalization establishes a relation of similarity among the entities it covers. A generalization entails that its instantiations are similar in certain respects. Consequently, adopting a system of laws involves a conceptual structuring of the pertinent realm of phenomena; in particular, it involves assumptions as to what is alike and what is not.

This means that incommensurability is a straightforward consequence of a semantical theory along with the historical observation that substantial theoretical revisions indeed occur. This suggests that incommensurability is real and instantiated. I defend this view by presenting an example and pinpointing the origin of non-translatability. I then reconstruct the emergence of incompatible structures of scientific kinds and finally analyze the bearing of incommensurability on the comparative evaluation of theories on empirical grounds. My claim is that this latter endeavor is not impaired in any serious respect: empirical comparison does not require translation and remains largely unaffected by incommensurability.

Kuhn adheres to the context theory of meaning (see Kuhn, 1983, p. 576–577; Kuhn, 1987, p. 8; Kuhn, 1989, p. 12, 15–20; see also Irzik and Grünberg, 1995, p. 297–298). According to this semantic theory, the meaning of a concept is determined by its relations to other concepts and the meaning of a statement results from its integration in a network of other statements. The conceptual and theoretical context provides the basis of meaning ascription. Another way of putting this is to say that the *use* of a concept determines its meaning. What a concept means is represented by the way in which it is applied to different situations. In the case of scientific concepts the usage is fixed by the laws of nature in which these concepts feature. This is to say, the meaning of a concept such as "electric field" is given by its lawful connections to related concepts such as "electric current", "charge" or "magnetic field". The concept "electric field" is understood if it is known, for instance, that such fields arise from electric currents or changing magnetic fields and produce alterations in the motion of charges, and so forth. Such relations between concepts imply connections among the truth values of statements. The given conceptual relations entail, for example, the truth of the statement that varying magnetic fields generate electric fields. Laws and theories supply a concept with a network of relations to other concepts, and they in this way determine to which situations the concept is to be applied appropriately. The theoretical context of a term fixes its use and determines its meaning.

Translation requires the coordination of a linguistic item with another one taken from a different language but possessing the same meaning. Adequate tanslation needs to preserve the meaning of the relevant concepts or statements. Within the framework of the context theory, this requirement of unchanged use is to be spelled out to the effect that two demands are to be fulfilled by translations. First, cognitive integration should coincide for the two items at issue. This applies, in particular, to the reproduction of standing inferential relations among statements. After all, it is such relations that provide the context relevant to the ascription of meaning. For example, the sentence "the tree over there loses its foliage" implies: "there is a deciduous tree". Analogously, "Wilfried is a bachelor" entails "Wilfried is not divorced". These logical relations should be preserved among the translated counterparts of these sentences. Second, the conditions of application of concepts should remain unaltered. One of the reasons why the German predicate *"hat schwarze Haare"* is disqualified as a translation of "is a bachelor" is that their conditions of application differ wildly. In fact, these conditions exhibit hardly any correlation. The two predicates are only accidentally applied to the same objects; they differ in meaning for this reason.

On the whole, then, the theoretical context account recognizes two chief determinants of meaning. First, cognitive integration of a concept or a statement which is given by the relations among the concept or statement at hand and other such items from the corresponding semantic field. The integration of scientific concepts or statements, in particular, is provided by the pertinent laws or theories. Second, conditions of application constitute the other pillar of semantically relevant use. Semantic properties are determined by the set of situations to which a concept is thought to apply. To these two sources of meaning correspond two constraints of adequate translations. Rendering a concept appropriately demands, first, the preservation of the relevant inferential relations among the pertinent statements, and, second, the retention of the conditions of application.

Owing chiefly to the investigations of Kuhn, it became clear that the historical development of science proceeds at least sometimes through stages of significant and deep-reaching conceptual and theoretical alteration. There is drastic theoretical change in science. This view finds its most prominent expression in Kuhn's characterization of scientific revolutions. Kuhnian revolutions are conceived as non-cumulative transitions. They do not involve the sustained elaboration and expansion of an accepted conceptual framework. On the contrary, a scientific revolution à la Kuhn consists in the revocation of fundamental principles of a discipline and their replacement by disparate ones. Furthermore, the disparity between pre- and post-revolutionary principles prohibits any smooth integration of the former into the framework of the latter. As a result of the far- reaching divergence between them, the pre-revolutionary theory cannot be reconstructed as the limiting case of the post-revolutionary one (see Kuhn, 1962, Chaps. VIII–X).

Kuhn initially located the chief origin of theoretical disagreement in revolutionary periods in perceptual, methodological, and ontological changes, but later distinguished linguistic deviation as the crucial feature of scientific revolutions. Incommensurability or non-translatability is shifted to center stage (see Kuhn, 1983, p. 684; Kuhn, 1987, p. 19–20). A theory is thought to comprise a "lexicon" which contains the kind terms employed by the relevant scientific community. These kind terms indicate what is assumed as being of the same kind; they represent expected similarity relations among objects or processes. Kind terms are connected to the laws of the corresponding theory. Incommensurability is said to arise from the non-translatability of kind terms, which is attributed, in turn, to changing expectations of similarity (see Kuhn, 1993, p. 315–318, 328, 336; Irzik and Grünberg, 1998, p. 211–212).

As indicated earlier, I refrain from addressing Kuhnian lexical structures more thoroughly and rather tackle the problem of incommensurability from a different, albeit related, angle. Kuhn recognizes that the structure of kinds is tied up with the generalizations that are constitutive of the corresponding theory (see Kuhn, 1987, p. 20; Kuhn, 1993, p. 316–317). So let me turn to the relations among laws from disparate theories and to the consequences that emerge with respect to the options for giving adequate translations. In order to fix ideas I proceed by elaborating an example. An often quoted instance of a Kuhnian revolution is the Einsteinian Revolution, a part of which involved the substitution of classical electrodynamics with special relativity. This is a revolutionary change by any measure so that the presence of incommensurable linguistic items can safely be expected – if incommensurability is supposed to be instantiated at all. I begin by giving a brief sketch of the relevant theories and subsequently point out the pertinent conceptual relations.

3 Shifting Theoretical Ground: the Example of the Einsteinian Revolution

Characteristic of classical, Newtonian mechanics is a principle of relativity that involves the equivalence of all frames of reference in uniform rectilinear motion with respect to mechanical processes. The so-called Galilean transformations connect spatiotemporal quantities obtained in one such inertial frame, S, to the quantities registered in another one, S', which is in inertial relative motion with respect to the first. If a given body is placed at some location, x, y, z, at some point in time, t, and is moved with a velocity, v, its spatiotemporal properties x', y', z', t', v', as measured within S', are specified by these Galilean transformations.

Before the advent of relativity theory, this equivalence of inertial frames was supposed not to extend to electrodynamics. Characteristic of Hendrik A. Lorentz's electrodynamic theory is the assumption of two kinds of entities, an "immovable" ether and electrically charged particles, the "ions" which later became the "electrons". These particles move in the all-pervasive ether. They are acted upon by what is now called the Lorentz force, which was thought to represent the force exerted by the ether on charged particles. The particles are capable of oscillations which were assumed to produce electromagnetic fields. The ether was held to embody an absolute rest frame in that all true motions appeared as motions with respect to the ether. The quantities figuring in electrodynamic equations bear witness to this interpretation. For instance, the velocity of light, as it features in Maxwell's equations, was viewed as the velocity relative to the ether. Likewise, current density, i.e., moved charges, which is also part of Maxwell's equations and of the Lorentz-force equation, was considered to involve absolute motion. In sum, the ether constitutes a unique frame of reference for Maxwell's and Lorentz's equations, and it exerts causal influences on charged objects that pass through it. In this way, the ether mediates the interaction between charged objects (see Schaffner, 1972, p. 113, 256; Nersessian, 1986, p. 211.)

Consequently, the equations of electrodynamics are restricted to frames at rest in the ether; they do not hold for moving systems without adaptation. In particular, the velocity of light, c, should depend on the motion of the observer. Lorentz attempted to cope with this problem by introducing his "theorem of corresponding states" which involved the distinction between universal time and local time. Universal time t is the correct measure of time; it refers to observers at rest in the ether. Local time t_L is dependent on the position x and the motion v of the system: $t_L = t - vx/c^2$. Local time was interpreted by Lorentz as a mathematical aid without direct empirical significance; it does not give the time readings of moved clocks (in contrast to its role in special relativity). In Lorentz, local time served the purpose of reducing a system of moved bodies to such a system at rest. The Galilean transformations entail unchanged time values for bodies irrespective of their states of motion. In Lorentz's approach, however, introducing an additional transformation by substituting universal time with the local time yields the correct electrodynamic quantities in moved frames. Assume a moved frame S in which electromagnetic fields are given at the spatiotemporal location x, y, z, t; then the values of these field quantities in the rest frame S_R can be obtained by taking their values in S at the same space coordinates and the corresponding local time x, y, z, t_L. The theorem of corresponding states achieved a reduction of the electromagnetic behavior of moving charged bodies to the properties of such bodies at rest. It entailed, in particular, that the measured velocity of light came out the same in the moving and the resting system (see Lorentz, 1899, p. 267; Drude, 1900, p. 472–474; Miller 1981, p. 32–40; Nersessian 1986, p. 217–219).

However, the theorem of corresponding states only held true for effects of the first order in v/c. By contrast, the experiment of Michelson and Morley (1887) had shown that no effect of the earth's motion was detectable to the order v^2/c^2. That is, the velocity of light was independent of the motion of the observer even to second-order effects. In order to cope with this anomaly, Lorentz (following Fitzgerald) introduced his contraction hypothesis. The dimensions of a moved body in the direction of motion shrink to such a degree that the changes in the velocity of light induced by the motion of the observer are precisely compensated – as the Michelson-Morley null result demands.

> "In order to explain the negative result of this experiment Fitzgerald and myself have supposed that, in consequence of the translation, the dimensions of the solid bodies serving to support the optical apparatus, are altered in a certain ratio." (Lorentz, 1899, p. 268)

As soon as the body is set in motion, its length is really reduced. The decrease is produced by the interaction between moved matter and the ether (Lorentz, 1899, p. 270; Drude, 1900, p. 482; Nersessian, 1986 p. 224). The resting ether compresses the body in passage through it. The consequence is that a moved observer measuring the dimensions of a body at rest in the ether should register a dilation or lengthening of the body. Lorentz's contraction effect is asymmetrical (see Schaffner, 1972, p. 113).

Lorentz succeeded in deriving the contraction from the principles of his theory by drawing on the additional assumption that the intermolecular forces that were supposed to hold a body together and were thus responsible for the body's dimensions transform like electromagnetic forces. Behind this assumption lies the – not implausible – notion that these intermolecular forces are similar in kind to electromagnetic forces or even of the same nature. The contraction hypothesis allows for the extension of the theorem of corresponding states to many second-order effects. However, Lorentz did not regard the theorem as a relativity principle. First, he explicitly did not rule out the existence of tangible effects of the motion of bodies through the ether. Only "many" electromagnetic phenomena were thought to appear in the same way irrespective of the observer's state of motion (see Schaffner, 1974, p. 48). Second, Lorentz's theory entailed that the frame of reference at rest in the ether was distinguished among the class of inertial frames in that it alone yields the true measures of lengths and velocities. The motion through the ether distorts spatiotemporal quantities. An optics textbook of the period put it this way:

> "Another way of explaining the negative results of Michelson's experiment has been proposed by Lorentz and Fitzgerald. These men assume that the length of a solid body depends upon its absolute motion in space." (Drude, 1900, p. 481).

On the other hand, the motion produces further effects that precisely compensate for the initial distortion – at least in "many" relevant phenomena. Lorentz's account involves a sort of conspiracy among different effects brought forth by the motion of charged bodies. These factors are so contrived as to cancel each other out, concealing in this way the true motion of bodies from the unbefitting curiosity of human observers.

Albert Einstein's special theory of relativity, by contrast, proceeds from the so-called special principle of relativity which says that all inertial frames are equivalent in every physical respect. The principle does not alone address mechanical phenomena but also includes electrodynamic processes. There is no privileged rest frame for electromagnetic phenomena either. It follows that there are no absolute velocities; all velocities are relative to other bodies or frames of reference. The second axiom Einstein cites is that of the constancy of the velocity of light, which says that the velocity of light is independent of the velocity of the light source. This constancy axiom follows from Maxwell's theory; it is a theorem of pre-relativistic electrodynamics. The special principle and the constancy axiom together imply the invariance of the velocity of light according to which this velocity assumes the same value for all inertially moved observers. The special principle of relativity entails that the inertial motion of a system of charged bodies or of an observer has no impact on electromagnetic processes. That is, Einstein abolished both the distorting influence of the motion and the counteracting factors. According to special relativity, all inertial frames are equivalent right from the start; their equivalence does not derive from the subsequent action of a compensating mechanism.

Special relativity likewise entails the contraction of moved bodies. In fact, Einstein's formula precisely agrees with Lorentz's; both give the same ratio of length reduction for a moved body. But in spite of their mathematical identity, the Lorentzian and Einsteinian equations differ in meaning. Their semantic difference is rooted in the divergent understanding of the quantity "v" featuring in the equations. In Lorentz it means "absolute velocity", that is, velocity of the relevant body with respect to the ether rest frame. But in Einstein there is no such rest frame. The velocity in question rather is the relative velocity between the body and the observer. This contrasting understanding has important ramifications. It follows, namely, that Einstein's contraction, contrary to Lorentz contraction proper, is reciprocal. Consider the following switch in perspective. Viewed from the angle of the moved and seemingly shortened body, the observer, previously assumed at rest but now regarded as moved, does not appear dilated but contracted as well. Further, when an observer is moved along with a body so that both are at relative rest, Lorentz contraction occurs, to be sure, but remains hidden because of an equal contraction of the measuring rods. In Einstein, by contrast, contraction is absent under these circumstances. These considerations suggest the existence of significant conceptual discrepancies behind the superficial, specious identity of the formulas. I take a closer look at the relevant conceptual relations in the following section.

4   Incommensurable Quantities in Classical Electrodynamics and Special Relativity

The conceptual structures of classical electrodynamics and special relativity are different to one another. The challenge is to identify seemingly analogous concepts and to explore if the relations among them are sufficiently tight for accepting one as a translation of the other. The spatiotemporal and dynamic measures in each theory constitute candidates for translation. Thus, the issue is whether counterparts for concepts such as length, duration, velocity, and mass can be specified in the two theories at hand. I leave temporal measures out of consideration. The reason is that Lorentz retained universal time but later learned from Einstein that it is his local time, rather than universal time, that the clock readings provide. Even after this recognition, however, Lorentz endeavored to stick to universal time and never reached a clear position in this matter. Unlike time dilation, Lorentz explicitly acknowledged length contraction, and his pertinent formula agrees mathematically with Einstein's (see Sect. 3).

The issue, then, is translation. As I argued earlier, adopting the context account of meaning places two demands on adequate translations. First, the theoretical integration or the inferential relations among concepts or sentences need to be preserved; second, the application conditions of concepts should be retained (see Sect. 2). Let's see how *prima-facie* translations fare in light of these requirements.

The first try includes focusing on conditions of application and to translating according to equality of measuring procedures: quantities that are determined empirically in the same way can be translated into one another. This applies to spatial intervals and mass values. Consider the situation of a Lorentzian observer at relative rest with the object of scrutiny. Since the observer may pass through the ether, one has to recognize that the measuring rods might be distorted and fail to yield correct values. But since the lengths to be determined shrink as well, the length ratios obtained are reliable. The same holds for length measurements using round-trip signal transmission time. The velocity of light was supposed to be altered by a possible motion of the observer, to be sure, but since the traversed spaces would change as well, the length ratios measured are trustworthy in any event. In special relativity, both the distortion and the counteraction are missing so that both rods and signal times may legitimately be employed. It is true, there are situations in which Lorentzian and Einsteinian observers will pass different judgments as to the reliability of specific measuring procedures (see below), but the majority of cases is of the sketched type. That is, classical electrodynamics and special relativity roughly agree on the acceptability of length measurements. Analogously, both accounts license the determination of mass with the aid of the balance or by analyzing collision processes (thereby drawing on momentum conservation). In this approach to translation, it is the empirical import of a concept that fixes its semantically relevant usage. Lorentz's concepts of length and mass are in accordance with Einstein's, since their empirical indications largely agree. The translation is governed by the principle that the conditions of application of the coordinated terms coincide.

The drawback is that the inferential relations fail to be preserved. As to the spatiotemporal quantities, the crucial divergence arises with respect to the interpretation of the relevant velocities. In Lorentz's contraction formula the significant quantity is the velocity between the moved body and the ether; in Einstein's mathematically identical equation the important magnitude is the body's motion with respect to an observer. The disparate conceptual integration of the seemingly identical concepts of length and velocity becomes conspicuous once the relevant types of situation, as they emerge in the Lorentzian framework, are reconsidered in Einsteinian terms.

(a) Lorentzian situation: body and observer equally at rest in the ether – no contraction.
Einsteinian reconsideration: body and observer at relative rest – no contraction.

(b) Lorentzian situation: body in absolute motion, observer at rest in the ether – contraction of the body.
Einsteinian reconsideration: body and observer in relative motion – reciprocal contraction.

(c) Lorentzian situation: observer in absolute motion, body at rest in the ether – contraction of the observer issuing in an apparent spatial dilation of the body.
Einsteinian reconsideration: body and observer in relative motion – reciprocal contraction.

(d) Lorentzian situation: both body and observer in equal absolute motion – shrinkage of both body and observer but no net effect due to compensation (Michelson-Morley situation).
Einsteinian reconsideration: body and observer at relative rest – no contraction.

The Lorentzian and Einsteinian approaches differ in judgment as to whether or not contraction occurs in a particular type of situation.

This distinction is tantamount to a shift in the theoretical integration of the concept of length. Consider a case of circular (but approximately rectilinear) motion such as the annual revolution of the earth around the sun. In Lorentzian terms, the change in the direction of motion entails that the body cannot be at rest in the ether all the time. This implies that contraction occurs. In special relativity, by contrast, this inferential relationship is severed. The assumption of a change in the direction of motion has no determinate consequences as to contraction. All depends on the choice of the frame of reference. Conversely, whereas in special relativity the introduction of such a frame in a particular state of motion is sufficient for implying judgments as to the occurrence of contraction, no unambiguous consequences follow in classical electrodynamics. In the latter framework absolute velocities are needed for this purpose so that the relativistic connection between relative motion and contraction is lost. I conclude that characteristic inferential relations for the concept of length are different in the two theories.

The first approach to translation was based on the rule of retaining the conditions of application of the terms at issue. The coordination between analogous terms was supposed to be forged by the equality of the empirical indications. Lengths are unanimously measured by relying on rods or signal transmission times; mass values are determined using a balance or collisions. On the adoption of this procedure, length and mass come out as (roughly) synonymous concepts in the two theories. However, this translation rule fails to reproduce the relevant inferential relations. It falls short of underwriting adequate translations for this reason.

The second approach was directed at the preservation of these inferential relations. The pursuit of this line amounts to explicating "velocity" and "contraction" the way I did earlier, namely, by outlining the role they play in Lorentz's electrodynamic theory or special relativity, respectively. Assuming the relativistic point of view, one would say that Lorentzian velocities are not relative velocities but refer to the motion of a body relative to the ether, and one would add that these absolute velocities are allegedly responsible for length contraction phenomena. However, adopting such a translation rule leads to a dramatic change in the conditions of application of the translated concepts. There is no such thing as absolute velocity or as ether-caused contraction (let alone the concomitant spatial dilation). If the Lorentzian concepts of velocity and length reduction are simply grafted upon relativity theory, the concepts become empty. They are no longer legitimately applied to any phenomenon. The retention of the inferential relations is purchased at the expense of losing the conditions of application. It does not provide an appropriate translation for this reason.

The same reasoning applies, *mutatis mutandis*, to mass. Lorentz favored the electromagnetic mass concept, popular at the period, according to which the inertia of bodies arises at least partially from the interaction between charges and fields. Mass is in part a derived quantity, not a primitive one. In this vein, Lorentz distinguished between "real" and "apparent" mass. Real mass is mechanical mass which invariantly characterizes a given body. Apparent mass is electromagnetic mass which refers to the increase in inertia a charged body undergoes as a result of its motion through the ether. A moving charge constitutes an electric current which generates a magnetic field. This field acts back on the charge and reduces the accelerating effect of external forces (see Miller 1981, p. 46–47). Application of Lorentz's theory implied that the total mass of charged particles, i.e., the sum of real and apparent mass, was dependent on the velocity of the particles. In fact, Lorentz formula was mathematically identical to the relativistic increase in mass. Accordingly, there are two relevant concepts of mass in Lorentz, namely, real mass and total mass. Both concepts possess *prima-facie* analogues in Einstein, namely, rest mass and relativistic mass. However, the correspondence turns out to be spurious on closer scrutiny.

At first sight, Lorentz's real or mechanical mass is tied to Einstein's rest mass. Both can be obtained by mechanical procedures; they remain invariant and are thus characteristic of a given body. Lorentz's total mass appears to correspond to relativistic mass. Both exhibit the same velocity dependence, and they equally govern the dynamic behavior of the pertinent body. Both are assumed to express a body's inertia. However, the influence of the divergent construal of the spatiotemporal quantities extends to the interpretation of mass as well. Namely, Lorentz's real mass is obtained empirically if the body at issue is at rest with respect to the ether. Einstein's rest mass, by contrast, is measured by an observer at rest relative to the body. The same holds analogously for total mass and relativistic mass, respectively. The former exhibits a dependence on the velocity relative to the ether, the latter on the velocity of the observer.

Taking the relevant empirical measures as a point of departure, therefore, one might be tempted to regard Einstein's rest mass as the translation of Lorentz's real mass, and Einstein's relativistic mass as the translation of Lorentz's total mass. But the theoretical integration of the two terms diverges considerably. Consider a charged body in motion through the ether. Lorentzians judge that its mass departs from its real value and that it does so for all observers including those moved at the same speed with the body. In relativity theory, by contrast, all that matters is the relative motion between the body and an observer, with the result that the observer moved with the body employs the body's rest mass. In Lorentzian terms the total mass is relevant for capturing this situation; in Einsteinian terms it is the rest mass – which thwarts the alleged analogy between real mass and rest mass. Conversely, take a body at rest in the ether as viewed from a moved observer. From a Lorentzian perspective, the dynamic behavior of the body is to be analyzed using its real mass whereas special relativity would invoke relativistic mass – which vitiates the alleged coordination between total mass and relativistic mass. This consideration shows that the supposedly intertranslatable concepts are used differently. The *prima-facie* coordination of terms breaks down in a large number of instances. The reason is that the inferential relations of the supposed analogues do not agree. For instance, the assumption of relative motion between body and observer rules out the invocation of rest mass; but nothing follows as to the use of real mass. The inferential relations of rest mass and mechanical mass are at variance with one another, and the same holds for relativistic mass and total mass.

Although the values obtained for real mass and rest mass as well as for total mass and relativistic mass, respectively, are in agreement for most of the empirically relevant situations to which these concepts are applied, they cannot be translated into one another. The reason is that the interpretation of these terms is different, and physics does not deal with pure numbers but with interpreted quantities. The translation fails in spite of the fact that the theoretical integration of the quantities looks the same mathematically. This similarity is deceptive, since these quantities are construed disparately. They are analogous mathematically but incongruous substantively.

The upshot is that the translation of concepts from disparate theories leaves one with the stark choice between two equally unacceptable alternatives. The first one is to translate according to the relevant conditions of application. That is, the two terms are applied under the same observable circumstances. The catch is that the statements formed by using these terms do not exhibit the same inferential relations. Adherents of relativity theory refuse the idea of an absolute rest frame; consequently, appending to the theory a concept to this effect would create an inconsistency (as Feyerabend mentioned as a general feature of incommensurability; see Feyerabend, 1962, p. 58–59). In fact, the missing of absolute velocities constitute by no means a simple gap in relativity theory which could be bridged by introducing such quantities. Rather, a move of this sort would contradict the principle of relativity, which rules out any distinguished frame of reference. Pursuit of this line would make relativity theory incoherent. The second option is to translate such that the inferential relations are retained. This amounts to giving a general description of ether-based electrodynamics. But the spatiotemporal and dynamic concepts specified in this framework are empty from a relativistic perspective so that the conditions of application are not preserved. Consequently, these concepts taken from Lorentzian electrodynamics are incommensurable with their *prima- facie* relativistic analogues.

5 Incommensurability, Split-up of Natural Kinds and Shifts in Reference

According to the context account of meaning, the laws of science contribute to the determination of the meaning of the concepts figuring in them. In addition to being influential on meaning, laws are important in another relevant respect, namely, in deciding about what is equal in kind. The laws induce a connection among the pertinent entities in that they appear as instances of the same set of laws. For example, it is a law that all protons exhibit the magnitude of the electron charge and a half-integer spin value. Consequently, all protons are of the same kind in that they equally display these properties. The law generates a "natural kind" (see Fodor, 1974, p. 101–102.) Natural kinds represent the relations of similarity or sameness in kind that are implicitly circumscribed by the relevant laws.

Kuhn features the shifts in the similarity relations as the chief distinction of incommensurability (see Sect. 2). However, it is unjustified to place these shifts at the top; rather, they follow immediately once it is realized that, first, incommensurability involves incompatible laws and, second, a collection of laws entails a structure of natural kinds. Adopting a set of laws inconsistent with the ones previously accepted implies a shift in the associated taxonomy of kinds. However, apart from mistaking a derived property for a fundamental one, Kuhn is quite right in emphasizing that the discrepancy between the taxonomies of natural kinds is the chief obstacle to translation (see Kuhn, 1983, p. 683; Hoyningen-Huene, 1989 p. 211–212). Incommensurable theories introduce partially overlapping classes of natural kinds. Such classes are torn into pieces, and the debris is reassembled to form novel, disparate classes. What was formerly considered as being of the same nature may be regarded as different afterward. Conversely, what was thought to be different in kind may be taken as conceptually unified in the new theory. It follows that two items which fall under the same category in one account are possibly to be expressed using different concepts in the other account. Conversely, two items labeled distinctly in one approach could be addressed uniformly in the other approach. Cross-classification of this sort vitiates translation (see Kuhn, 1990, p. 4; Irzik and Grünberg, 1995, p. 299).

This reshuffling of similarity relations is evidenced by the list of Einsteinian reconsiderations of Lorentzian types of situations (see Sect. 4). Ties of similarity are unraveled, and others are established in their stead. For instance, from a Lorentzian perspective, a body in absolute motion observed from a frame of reference attached to the ether (situation (b)) is of the same type as the Michelson–Morley situation (situation (d)). Both are equally characterized by the occurrence of contraction. From Einstein's point of view, by contrast, the two situations are distinct in kind in that relative motion, and consequently contraction, is present in the former, but not in the latter. Conversely, in Einstein's framework all situations that involve the relative motion between body and observer are of the same type (situations (b) and (c)). By contrast, against the backdrop of Lorentz's theory, it makes a difference whether the body or the observer is in motion. The observer will register a contraction in the former case but a dilation in the latter. Disparate systems of laws tie different properties together; conflicting theories go along with collecting nomologically relevant properties into incompatible equivalence classes.

The same reasoning applies to mass. In the Lorentzian framework, situations of the same type are characterized by equal velocity with respect to the ether – irrespective of the motion of the observer. Such Lorentzian equivalence classes are dissolved in relativity theory through the introduction of the motion relative to an observer as the salient quantity. Conversely, Einstein forged other equivalence classes instead. Situations of the same type are characterized by equal velocity with respect to an observer – irrespective of the motion relative to the ether. That is, what was formerly conceptually united crumbled into separate pieces, and what was considered distinct previously was integrated conceptually.

The same goes for the relation between Newtonian and Einsteinian mass which is among the most frequently cited examples of incommensurable concepts. Newtonian mass is distinguished by the double role of, first, yielding an invariant characteristic of a given body and, second, determining the dynamic behavior of the body in physical interactions. These two features are separated in special relativity. The body is characterized by its rest mass, while its dynamic behavior is determined by its relativistic mass (see Sect. 4). So, what was considered a unique quantity in the framework of classical mechanics is split up into two in special relativity. But the converse is also true. In classical mechanics, kinetic energy is a separate quantity that is also influential on the dynamic behavior of a body. In special relativity, however, this influence is integrated into relativistic mass. Mass and energy are of the same kind. What were traditionally conceived as distinct factors are conceptually unified in relativity theory.[1]

---

[1] In his early analysis of the relation between the Newtonian and Einsteinian concepts of mass, Kuhn features the variability of relativistic mass and contrasts it with the conservation of mechanical mass (see Kuhn, 1962, p.101–102). But

Incommensurable concepts in science emerge if, owing to nomological change, natural kinds are restructured. In the course of adopting a system of laws in conflict with the previous one, the former equivalence classes split up into heterogeneous components and realign to form new taxonomic structures. The conceptual integration of a concept is determined, in large measure, by the collection of properties that fall under the concept. Changing this collection involves a change in the integration of the concept and the context relevant to its meaning. The inferential relations are replaced by others so that translation is vitiated. Consequently, the translation failure ultimately goes back to the contrast between two sets of laws. This contrast leads to the dissection and new formation of natural kinds, which is the proximate reason for the non-translatability of the corresponding concepts.

6   Empirical Comparison of Theories with Incommensurable Concepts

Meaning discrepancy due to nomological divergence has important ramifications. It is only due to this wider impact that incommensurability could be viewed as a threat to scientific rationality in the first place. Otherwise it might appear a quite plausible and utterly unexciting feature that the concepts of mistaken theories cannot be translated into the framework of their more correct successors. These ill-conceived concepts are dropped as the pertinent theories are superseded by improved approaches. The occurrence of a translation failure of the kind in question indicates that science progresses profoundly. For instance, it was realized by Einstein that there is no such thing as motion with respect to the ether. Therefore, it is only natural that a concept which was discovered to be misleading and empty cannot be integrated into the superior theory.

However, the more serious implications of this translational rift are realized once the semantic principle that meaning determines reference is taken into account. This principle is endorsed by a number of linguistic approaches

---

this trait would be insufficient for establishing incommensurability in the sense of the later Kuhn. It would rather suggest the identification of mechanical mass with relativistic rest mass. Feyerabend's treatment of the relation is less than convincing as well. As he argues, the relativistic concepts of mass and length are two-place predicates in that they express a relation between measured values and a frame of reference. The seemingly analogous concepts of classical mechanics, by contrast, are one-place predicates which lack any reference to such a frame. Consequently, from the relativistic point of view, these concepts are meaningless (see Feyerabend, 1970, p. 221–222; see also Sect. 6). But, in fact, all the relevant prerelativistic concepts involved a reference to such a frame, namely a preferred rest frame. Newton construed absolute motion as motion relative to absolute space; Lorentz regarded true motion as motion relative to the ether. Thus, the relevant concepts were considered two-place predicates throughout.

(e.g., it is a theorem of possible world semantics). Given that meaning determines reference, a change in the former suggests a shift in the latter as well. This is not a strict implication, to be sure, since it is possible that concepts with different meaning pick the same referents. However, in the large majority of cases such concepts refer to distinct objects or processes. Consequently, incommensurable concepts are likely to differ in reference. Theory change probably goes along with reference change so that the successor theory says different things about different objects – rather than different things about the same objects. This militates against a cumulative view of scientific progress. Progress cannot be regarded as being tantamount to understanding more and more aspects of the same entities.[2]

This might be considered bad news for traditional conceptions of scientific progress. But in fact matters are even worse. Incommensurability appears to threaten, in addition, the comparability of the *empirical* achievements of rivaling theories. Without a common referential ground, no overlap among the empirical findings and problems can be identified – or so it seems. Incommensurability appears to rule out any agreement as to which problems are to be solved and which facts are relevant. Actually, it is not even clear whether the theories at hand are competing with one another. A competition requires equality of reference; competing theories entail deviant predictions about the same objects. However, in case of incommensurable theories, as the argument runs, equality of reference can never be ascertained.

The strand of reasoning leading from translation failure to the exclusion of empirical comparison roughly looks as follows: Non-translatability implies that the claims of one theory cannot be expressed within the framework of the other and vice versa. It follows that the content of one theory cannot be captured by the other. But if what the allegedly rival theory says remains opaque, there is no way to judge if it agrees or contrasts with one's own theoretical assumptions. Consequently, no empirical comparison between the claims at issue seems possible. In this interpretation, incommensurability would not alone thwart the intertranslation of the cognitive content of the theories at hand but also vitiate their comparative empirical evaluation.

Feyerabend mentions the case of experiments that are supposed to discriminate empirically between special relativity theory and classical electrodynamics. An example is the series of experiments conducted by Walter Kaufmann in the period 1901 to 1905, which, considered with hindsight, were relevant for the relativistic dependence of mass on velocity – although they issued

---

[2] It is true, the so-called causal theory of reference (which grew out of Kripke's theory of proper names) can be invoked so as to avoid this consequence. The causal theory separates meaning from reference (and thus rejects the mentioned semantic principle). Consequently, on this account, a change in meaning does not imply a change in reference. But the causal theory has problems of its own (which I cannot address here). To mention just one, the causal account is at a loss to convincingly explain reference failure (see Cummiskey, 1992).

in the erroneous refutation of this dependence (see Hon, 1995, p.194–195; Cushing, 1998, pp. 210–215). However, as Feyerabend argues, mass in classical mechanics is an invariant property of a body, whereas in relativity theory it involves a relation between a body and an observer's frame of reference. Simply ascribing a mass value to a body makes no sense in relativity theory. Classical mass and relativistic mass are governed by disparate laws and are incommensurable for this reason. Consequently, the results of Kaufmann's experiments are to be described differently in either theory. The classical and the relativistic descriptions need to employ the classical and relativistic concepts of mass, respectively. But in view of the incommensurability of these concepts, it appears illicit to maintain that it is the "same experiment" that undermined relativity theory and confirmed traditional electrodynamics. It is possible, to be sure, to point to Kaufmann's apparatus and to say that the experiment performed with "this" device had the effects mentioned. But pointing to an apparatus falls short of characterizing an experiment. Experiments refer to *types* of activity; otherwise it would make no sense to speak of a repetition of the *same* experiment. This implies that experiments need to be identified through a descriptive characterization. But if incommensurable concepts are involved, no theory-neutral description can be given. Kaufmann's experiments in relativistic description are different from Kaufmann's experiments in classical description, since each description uses a concept of mass that is inexplicable in the framework of the other.[3].

Note that this account does not rule out empirical tests of a theory. It would still be possible to detect that special relativity contradicted the results of Kaufmann's experiments interpreted relativistically – as Feyerabend is prepared to admit (see Feyerabend, 1970, p. 226; Feyerabend 1975, p. 282). What would be prohibited, at most, is to say that classical electrodynamics was in agreement with the *same* results that served to discredit relativity theory.

But in fact, it is in no way ruled out generally to empirically compare theoretical claims couched in incommensurable concepts. The point is that incommensurable theories need to be comparable in some respect in order to generate a non-trivial translation problem in the first place. A host of theories is not translatable into one another without anything significant coming out of it. Darwin's theory of natural selection is not translatable into hydrodynamics; quantum mechanics cannot be rendered by the concepts of Zen. In order for non- translatability to become a significant issue at all, such cases need to be excluded. The obvious way to do this is to draw on one of the defining features of incommensurability, namely, inconsistency of the laws involved. Incommensurable concepts are not translatable since the rele-

---

[3] Although Feyerabend proceeds toward this position, he does not fully embrace it eventually; see Feyerabend, 1970, p. 220–222, 226 [see also the more poignant formulation in the German version Feyerabend, 1978, p. 184–185, 190]; Feyerabend, 1975, p. 282–283

vant laws, as specified within each of the theories at hand, are incompatible with one another (see Sect. 2). Feyerabend distinguishes between competing and independent theories and restricts incommensurability to concepts from theories of the former kind (see Feyerabend, 1972, p. 304). But no such inconsistency occurs in one of the just-mentioned examples.

The salient point is that a conflict between two theories only emerges if there is some shared realm which they jointly address. The significance and non- triviality of incommensurability requires that there is some range of phenomena that can be considered relevant for both theories (see Hoyningen-Huene, 1989, p. 213). This common ground is sufficient to enable one to compare some of the empirical consequences of the theories involved. Consider a particular experiment for which both theories claim responsibility. Each of them captures the outcome by using its own observational vocabulary. In such a situation an empirical comparison between the two approaches is feasible by using exclusively concepts taken from the pertinent approach. Assume that the experimental result is such that it is justified, on the basis of one account, to conclude that a certain process, as specified in terms of this account, has actually taken place, while it is illicit, on the basis of the rivaling theory, to judge that the process required by this theory has occurred. That is, the outcome of the experiment is framed by exclusive invocation of the terms of one's own theory. Still, the phenomenon is addressed by both theories; thus, it is the *same* experiment whose outcome is reported.

Kaufmann's experiments constitute an example. They were directed at the empirical determination of the dependence of the electron mass on the particle's velocity. But mass in Lorentz and mass in Einstein are incommensurable quantities (see Sect. 5). Actually, purpose and theoretical impact of the experiments differ wildly according to the point of view one is inclined to take. Kaufmann's intention was to measure the ratio of real and apparent mass, and his conclusion was that the electron is devoid of any mechanical mass. Its inertia is entirely of electromagnetic origin and nature. In particular, Kaufmann thought he had buttressed the account of Max Abraham, which proceeded on the assumption of a "rigid" (non-deformable) electron of vanishing mechanical mass. Kaufmann supposed he had undermined Lorentz's theory of the deformable electron which arose from the application of the Lorentz contraction to the moved electron. Lorentz's theory predicted the same velocity dependence as special relativity so that the latter was considered refuted as well. This time no specific model of the electron was at stake but rather the principle of relativity (see Hon, 1995, p. 184–189; 194–197).

The significance of the quantities obtained and the envisaged theoretical impact of the experiments was markedly at variance in the three approaches under consideration. Still, the finding was held relevant by all of them. The measurement was done by registering electron trajectories in electromagnetic fields. The electron paths were detected using a photographic plate. Kaufmann managed to obtain a visible curve on the plate, whose precise shape was supposed to indicate the sought-after quantity. This effect was positively identified irrespective of the theoretical background. The reliability of the result was contentious, to be sure; actually, it was abandoned later. But this issue had nothing to do with the dissent as to the overarching background principles. Adherents of the rigid electron, defenders of the deformable electron, and relativists experienced no trouble whatsoever in unanimously identifying the traces left by incident electrons.

Another case in point is the Kennedy-Thorndike experiment of 1932, which constituted a refinement of the Michelson-Morley experiment and was undertaken so as to compare Lorentz's and Einstein's theories empirically. Lorentz's account entailed a negative result of the Michelson-Morley experiment only on condition that the lengths of the interferometer arms are equal (when unaffected by the motion through the ether). Only under such circumstances did the changes in the velocity of light induced by the motion of the observer and the contraction of the optical apparatus cancel each other out precisely. Interferometers equipped with unequal arms, by contrast, should exhibit a net effect. Electrodynamic theory yields an equation that connects the ensuing shift in the interference fringes with the difference in the lengths of the interferometer arms (resting in the ether) and the velocity of the apparatus with respect to the ether. The prediction was that if these two quantities did not vanish, a shift should occur during the seasonal change in the direction of the earth's motion. The reason (as given above) is that whatever the precise movement of the earth through the ether it cannot be at rest all the time (see French, 1968, Sects. 3.1 and 3.6).

The partisan of classical electrodynamics may feel free to determine the relevant quantities in whichever way she prefers. The point is that the circumstances can easily be arranged such that her theory entails the appearance of fringe shifts. The only thing she has to acknowledge is the inequality of the interferometer dimensions and the attribution of a non-vanishing absolute velocity to the earth. The realization of the former condition can be left completely to her; the fulfillment of the latter follows from her theory. By contrast, the adherent of special relativity anticipates that no fringe shifts occur. From his point of view, the Kennedy-Thorndike experiment is but a trivial modification of the Michelson- Morley experiment, so the same null result is to be expected. As it turned out, the prediction of the latter was confirmed, and the expectation of the former disappointed.

In this case, one theory was successful on the very turf where the other was defeated. And at least one such range of common relevance has to exist so as to create non-trivial incommensurability in the first place. Consequently, all pairs of incommensurable theories necessarily possess at least one realm of phenomena which they jointly address and which provides the basis for their.empirical comparison (see Papineau, 1979, p. 137–138; see also Laudan, 1977, p. 142–144).

The crucial aspect is that the success or failure of the empirical test is judged against the background of one's own commitments and standards. No need for translation arises. The reasoning from translation failure to empirical non-comparability proceeds by arguing that claims made using inexplicable concepts cannot be understood and, consequently, cannot be put to empirical scrutiny (see above). But it is in no way mandatory to have the empirical claims of one theory checked by the adherents of another theory. A theory may well be tested by its own followers, and to them the relevant claims are by no means obscure. The only thing necessary to proceed from empirical test to empirical comparison is a shared realm of relevant phenomena. Advocates of each theory have to acknowledge responsibility for coping with these phenomena (which may be disparately understood in either theory). This much of a common ground is secured by the mere fact that we are dealing with incommensurable theories.

Theories can be compared empirically without having to translate their theoretical principles into one another. The standards for appropriate translation are harder to satisfy than the requirements for empirical comparison. Empirical comparison demands that instantiations of observational consequences can be correlated, whereas translation requires the mapping of theoretical concepts under the joint preservation of their inferential relations and their conditions of application (see Sect. 2). For this reason, empirical comparison need not be impaired by the shift in reference and the restructuring of natural kinds that is characteristic of incommensurable theories. All that is needed is the identification of an experiment or phenomenon as lying within the domains of application of both theories involved. Only if this much of a common ground exists between the theories can any conflict or inconsistency between their laws arise. Theories addressing completely disjunct sets of phenomena are compatible with one another and hence cannot be incommensurable. It is precisely the amount of shared features which makes translation failure non-trivial in the first place that secures the possibility of empirical comparison.

It is true, the range of empirical overlap between incommensurable theories might be narrow and insufficient for an unambiguous comparative evaluation. But problems of that sort may arise for each pair of theories. There is no guarantee that in comparing two theories one comes out first distinctly. But uncertainties of this kind haunt empirical comparison in general and have nothing specifically to do with incommensurability. The upshot is that theories containing incommensurable concepts can be evaluated with regard to their comparative empirical achievements – albeit subject to those constraints that restrict empirical comparison in general.

7   Conclusion

These considerations suggest that incommensurability constitutes an immediate consequence of a particular semantic theory, namely, the context account of meaning or semantic holism. This is borne out by the fact that Carnap, starting off from a similar premise, arrived at a similar conclusion. Carnap adopted a sort of context theory by claiming that theoretical postulates serve the double purpose of determining partially the meaning of theoretical terms and of expressing factual content. He inferred that concepts taken from different domains of science (he mentioned classical and modern physics) may turn out to be untranslatable into one another. That is, divergent theoretical approaches engender non- translatability among their concepts (see Irzik and Grünberg, 1995, p. 290–291).

Theoretical disparity and semantic incommensurability are related to one another in a straightforward way. The fact that this relationship is acknowledged and endorsed by Carnap made it appear less than convincing to regard incommensurability as the hallmark of irrationality – as is frequently claimed.[4]. This irrationality claim seems even less plausible in light of the result presented above, namely, that the objectivity of science or its commitment to experience has nothing to fear from incommensurability (see Sect. 6).

Finally, incommensurability continues to be of epistemic significance in one respect. It contributes to undermining a cumulative view of scientific progress according to which science manages incessantly to pile up truths upon one another. The lesson incommensurability teaches is that losses occur as well. In the course of theory change, some scientific achievements are conceptually reframed beyond recognition. In particular, the occurrence of reference shifts poses a serious threat to the claim that scientific theories accomplish an ever deeper understanding of the same objects and processes. Actually, one of the targets Kuhn and Feyerabend had aimed at by introducing the argument from incommensurability was to overthrow the cumulative view of scientific progress.[5]. In this respect the incommensurability thesis retains some force after all. It contributes to undermining convergent realism according to which the succession of theories in the mature science is to be viewed as an approach to the true account of nature. Consequently, the qualms raised initially as to the representational power of theories are not mitigated by the analysis of incommensurability. They cannot be laid to rest.

This result underlines the creative aspect of concept formation in science. The symbolic nature of scientific concepts makes them dependent on human

---

[4] The frailty of this allegation is emphasized by the fact that Carnap was among the editors of the series in which Kuhn (1962) appeared and approved the acceptance of Kuhn's work for publication (see Salmon, 1999, p. 347).

[5] See Feyerabend, 1970, p. 219; Kuhn, 1993, p. 330. Kuhn's anti-cumulative approach is restricted to revolutionary periods (in which incommensurability is thought to occur); normal science, by contrast, is considered a "cumulative enterprise" (see Kuhn, 1962, p. 52).

epistemic powers and purposes. Each theoretical shift involving incommensurable concepts suggests that there is some way to go until scientific concepts fit nature like a glove.

References

Carrier, M. (1994): *The Completeness of Scientific Theories. On the Derivation of Empirical Indicators within a Theoretical Framework: The Case of Physical Geometry (Western Ontario Series in the Philosophy of Science 53)*, (Kluwer, Dordrecht).

Cummiskey, D. (1992): "Reference Failure and Scientific Realism" in *Brit. J. Philos. Sci. 43*, 21–40.

Cushing, J. T. (1998): *Philosophical Concepts in Physics*, (Cambridge University Press, Cambridge).

Drude, P. (1900): *The Theory of Optics*, (Dover Publications, New York, 1959).

Feyerabend, P. K. (1962): "Explanation, Reduction and Empiricism," in Feigl, H. & Maxwell, G. (Eds.): *Scientific Explanation, Space and Time*, (University of Minnesota Press, Minneapolis), 28–97.

Feyerabend, P. K. (1970): "Consolations for the Specialist" in Lakatos, I. & Musgrave, A. (Eds.): *Criticism and the Growth of Knowldege* (Cambridge University Press, Cambridge), 197–230.

Feyerabend, P. K. (1972): "Die Wissenschaftstheorie – - eine bisher unerforschte Form des Irrsinns?" in Feyerabend (1978a), 293–338.

Feyerabend, P. K. (1975): *Against Method. Outline of an Anarchistic Theory of Knowledge* (Verso, London), 1975, [7]1978.

Feyerabend, P. K. (1978): "Kuhns Struktur wissenschaftlicher Revolutionen: Ein Trostbüchlein für Spezialisten?" in Feyerabend (1978a), 153–204 (revised version of Feyerabend [1970]).

Feyerabend, P. K. (1978a): *Der wissenschaftstheoretische Realismus und die Autorität der Wissenschaften (Ausgewählte Schriften I)*, (Vieweg, Braunschweig, 1978).

Fodor, J. A. (1974): "Special Sciences (or: The Disunity of Science as a Working Hypothesis)," *Synthese 28*, 97– 115.

French, A. P. (1968): *Special Relativity*, (MIT Press, Cambridge Mass.)

Hempel, C. G. (1966): *Philosophy of Natural Science*, (Prentice Hall, Englewood Cliffs N. J.).

Hon, G. (1995): "Is the Identification of Experimental Error Contextually Dependent? The Case of Kaufmann's Experiment and Its Varied Reception," in Buchwald, J. Z. (Ed.): *Scientific Practice. Theories and Stories of Doing Physics*, (University of Chicago Press, Chicago) 170–223.

Hoyningen-Huene, P. (1989): *Die Wissenschaftsphilosophie Thomas S. Kuhns*, (Vieweg, Braunschweig).

Irzik, G. & Grünberg, T. (1995): "Carnap and Kuhn: Arch Enemies or Close Allies", *Brit. J. Philos. Sci. 46*, 285–307.

Irzik, G & Grünberg, T. (1998): "Whorfian Variations on Kantian Themes: Kuhn's Linguistic Turn", *Stud. Hist. Philos. Sci. 29*, 207-221.

Kuhn, T. S. (1962): *The Structure of Scientific Revolutions*, University of Chicago Press, Chicago, [2]1970.

Kuhn, T. S. (1983): "Commensurability, Comparability, Communicability", *PSA 1982. II*, (PSA, East Lansing Mich.), 669–688.

Kuhn, T. S. (1987): "What are Scientific Revolutions?", in Krüger, L. et al. (Eds.): *The Probabilistic Revolution I: Ideas in History*, (MIT Press, Cambridge Mass.), 7–22.

Kuhn, T. S. (1989): "Possible Worlds in History of Science", in Allén, S. (Ed.): *Possible Worlds in Humanities, Arts and Sciences*, (de Gruyter, Berlin, 1989) 9–32.

Kuhn, T. S. (1990): "The Road since Structure", *PSA 1990.II*, (PSA, East Lansing Mich.), 3-13.

Kuhn, T. S. (1993): "Afterwords", in Horwich, P. (Ed.): *World Changes: Thomas Kuhn and the Nature of Science*, (MIT Press, Cambridge Mass.) 311–341.

Laudan, L. (1977): *Progress and its Problems. Toward a Theory of Scientific Growth*, (University of California Press, Berkeley).

Lorentz, H. A. (1899): "Simplified Theory of Electrical and Optical Phenomena in Moving Systems", in Schaffner (1972), 255–273.

Papineau, D. (1979): *Theory and Meaning*, (Clarendon Press, Oxford).

Miller, A. I. (1981): *Albert Einstein's Special Theory of Relativity. Emergence (1905) and Early Interpretation (1905-1911)*, (Addison-Wesley, Reading Mass.).

Nersessian, N. J. (1986): "'Why Wasn't Lorentz Einstein?' An Examination of the Scientific Method of H.A. Lorentz," *Centaurus 29*, 205–242.

Salmon, W. (1999): "The Spirit of Logical Empiricism", *Philos. Sci. 66*, 333–350.

Schaffner, K. (1972): *Nineteenth-Century Aether Theories*, (Pergamon Press, Oxford).

Schaffner, K. (1974): "Einstein versus Lorentz: Research Programmes and the Logic of Comparative Theory Evaluation", *Brit. J. Philos. Sci. 25*, 45–78.

7. Symbol and Intuition in Modern Physics

Brigitte Falkenburg

Physical theories are said to be non-intuitive, for they are formulated in the abstract and symbolic language of mathematics. At the same time, they are supposed to model real things, events or processes, that is, they are supposed to furnish ideal descriptions of actual states of applying inside or outside of a physical laboratory. Today, it is frequently suggested that it was the scientific revolutions of the 20th century which first presented us with abstract and non-intuitive theories and which led to serious problems of interpretation that shook our traditional understanding of reality. According to the classical conception, the concrete and intuitive meaning[1] of physical theories and models counts as the necessary condition of their reference. Kant's theory of nature in particular ascribes to intuition the function of filling the semantic gap between the formal language of mathematical physics and the characterization of its empirical domain of objects. The physics of the 20th century, by contrast, developed theories and models for the world on a large as well as on a small scale which no longer correspond to the classical ideas. The special and general theories of relativity as well as the quantum-mechanical description of reality are compatible with an intuitive representation of their theoretical approaches and specific models only to a very limited extent.

Today's *common sense* regarding what counts as intuitive, however, is a product of historical processes. It was formed only long after the development of modern physics – not least due to Kant's theory of intuition. For an Aristotelian of the 17th century, the Copernican theory and Newtonian mechanics were just as non-intuitive as the theories of Planck, Einstein, Bohr, Heisenberg, Schrödinger, Weinberg or Hawking are for us today, theories which are no longer compatible with the cosmological ideas and the understanding of reality of the 18th and 19th centuries. In what follows, I would first like to sketch the traditional epistemological debates concerning the language and reality of physics (Sect. 1) and then proceed to show how Kant's theory of intuition was supposed to close certain semantic gaps between the language of mathematical physics and our world of experience (Sect. 2). Finally, I want to indicate, with recourse to Kant, to what extent contemporary physics too relies on intuitive concepts and models in order to embed its abstract system descriptions into natural language usage (Sect. 3).

---

[1] Translator's note: here and in the following, "meaning" stands for the general German expression "Bedeutung", whereas "reference" stands for Frege's "Bedeutung", as opposed to "sense" ("Sinn"). Translated by Hans-Jakob Wilhelm.

1   Language and Reality

The mathematical and experimental procedure of physics leads us away from our ordinary understanding of reality, away from the usual qualitative experience of the events occurring around us and away from the natural language in which we normally express our experiences. The experiments of physics are designed to generate phenomena which can be classified and turned into the object of mathematical physics. The book of nature, which, according to Galileo, is written in mathematical characters, does not simply lie open before us, save perhaps for the starry sky. In large parts, we must first write the book into the phenomena by means of technical devices before we can begin to decipher it. Galileo's experimental method is constitutive for all of modern natural science – for physics, chemistry, biochemistry, microbiology as much as for the application of these disciplines in gene technology, medicine or the geosciences. Its aim is to isolate certain partial aspects of natural phenomena which are analyzed under the most ideal natural and technical conditions possible. The experiments of physics and of the other modern natural sciences are designed to produce regularized and reproducible phenomena and to vary them in a controllable manner. In order to achieve this, one must already have theoretical knowledge or at least specific expectations regarding the natural phenomena and their concurrence. Every experiment is designed to decompose complex natural events into specific components, which are then systematically investigated in order to establish numerical values and functional connections for physical quantities. Only the disassembly of phenomena into regularized components – the analytic-synthetic procedure according to the experimental method of Galileo – makes the application of mathematics to natural phenomena possible. Only an observable effect in an experimental setup, one that can be arbitrarily reproduced in a controlled manner, can be grasped as an element of a well-defined class of homogeneous physical phenomena which then become accessible to mathematical description.

Thus, the experimental method serves the generation of well-defined classes of phenomena and the investigation of the systematic relations of their elements. And hence the results of experiments correspond to the abstract and symbolic character of physical laws and theories which was especially emphasized by Pierre Duhem.[2] Physical laws are expressed by means of mathematical symbols such as $x, m, \frac{dx}{dt}, p, K$ and q, which stand in certain functional connections such as $p = m\frac{dx}{dt}$. The formal meaning of these symbols is implicitly defined by the axiomatic basis of the theory in which they occur – through axioms such as the law of force, $K = \frac{d^2x}{dt^2}$, of classical mechanics. The physical interpretation of these formal symbols then occurs through concepts of a quantity such as location, mass, velocity, force, energy, charge or tem-

---

[2] See his analyses of the relation between theory and experiment in *Ziel und Struktur physikalischer Theorien* (Duhem 1906).

perature. The physical meaning of a concept of a quantity such as "mass" is abstract. It does not reside in concrete objects, but in a class of physical properties with a whole scale of corresponding numerical values. Every measured value in turn corresponds to a class of concrete phenomena which can be produced under well-defined experimental conditions. Thus, following the terminology of Frege's semantics,[3] the concepts of a quantity in physics are second-order concepts:

1. Every numerical value of a physical quantity corresponds to a class of concrete physical phenomena which can be produced in experiments under well-defined conditions and for which a certain type of measuring device or procedure delivers, within certain margins of error, one and the same measured value.
2. A quantity is formally defined as a function which maps a class of properties onto a set of real numbers[4] such that a determinate quantitative value corresponds to each real number. This set of real numbers – the scale of a quantity – is defined only up to the point of the choice of a unit of measurement and generally ranges from 0 to ∞.

Thus, every concept of a quantity in physics refers to a class of classes of concrete physical phenomena – or to a class of physical properties constituting a scale of numerical values and ascribable to concrete phenomena which can be generated under well-defined experimental conditions. In this regard, most physicists hold the view that the formation of such classes in physics rests on fundamental physical properties which belong to entities such as atoms, elementary particles or black holes which are not immediately observable. This is an *essentialist* position which holds that the concepts of a quantity in physics aim at essential properties (or primary qualities) of natural phenomena. The class of properties corresponding to a concept of a physical quantity, however, can also be defined operationally – without any essentialist metaphysics – through a connection to measurement processes. This requires the definition of a chain of empirical measurements by means of which the scale of a quantity can be completely apprehended.[5]

Duhem was *not* an essentialist. He defended an anti-metaphysical interpretation of the abstract and symbolic language of physical quantities. According

---

[3] See Frege's essays on semantics (Frege 1892a,b), as well as his definition of the concept of a number as a second-order concept (Frege 1884).

[4] This applies only to classical physics with the inclusion of both theories of relativity; in a quantum theory, the quantities are not real-valued, but operator-valued. The classical concept of a quantity derives from Newton. According to him, the relation between two arbitrary (empirical) quantities can be expressed by means of a real number; see the article *Größe* by Suppes in the German dictionary *Handbuch wissenschaftstheoretischer Grundbegriffe* (Suppes, 1980).

[5] See my essay *Incommensurability and Measurement* (Falkenburg, 1997). According to Bridgman's radical operationalism, by contrast, every measurement method defines a distinct type of quantity.

to him, physical concepts, laws and theories do not stand for things or events *in concreto*. Neither did he regard them as elements of a true description of a physical reality which would lie at the basis of observable phenomena in the form of essential properties and true causes.[6] From Duhem's perspective, the abstract concepts of a physical theory are mere *instruments*. He emphasized that their definition always includes a certain measure of arbitrariness and that they serve primarily to bundle, in the most economical way, as many qualitatively most distinct phenomena as possible. On the first issue, he tended towards Poincaré's conventionalist view of physics, according to which there always remains an amount of arbitrariness regarding the manner in which concepts of physical quantities are defined and physical laws are formulated in agreement with experimental results. On the second issue, Duhem was in close proximity to Ernst Mach's empiricist view of physical theory. According to Mach, physical theories only aim at an economical representation of experimental phenomena.

The founders of contemporary physics, by contrast, defended a scientific realism, according to which physical quantities and theories aim at the essential properties and structural characteristics of things and events in nature. Boltzmann, Planck, Einstein, Rutherford and Bohr were atomists, personally participating decisively in the investigation of atoms. In his lecture, *Die Einheit des physikalischen Weltbildes*, of 1908, Max Planck distanced himself decidedly from Mach's empiricist and phenomenalist conception of physics and presented a realist and essentialist, if not Platonist, view of the abstract symbols of physical theory. According to Planck, the formation of physical concepts aims at liberating our understanding of nature more and more from anthropomorphic conceptions. Thus, the development of the concept of force of classical mechanics emancipated us from the idea of the bodily force which we must apply in order to do work, for example, when we want to lift an object.[7] Contrary to Mach, Planck does not regard the increasing distance of physical theory from immediate sense experience as a loss, but rather as a gain – or, stated more adequately, the gain, in his eyes, by far exceeds the associated loss:

> "If we look back upon the past, we can briefly summarize by saying that the signature of the past development of theoretical physics is a unification of its system which is achieved through a certain emancipation from the anthropomorphous elements, especially sensations. ... Indeed, the advantages must be invaluable, if they deserve such fundamental self-sacrifice!" (Planck, 1965, p. 31)

The "invaluable advantages" lie, as Planck subsequently explains, in the increasing unity of the physical view of the world. For him this means much

---

[6] In the spirit of Newton's first rule of reasoning in philosophy (Newton 1729) p. 398.

[7] Planck (1965), p. 30.

more than an economy of thought in Mach's sense; he emphasizes that the unification of theories leads to a physical *universalism*. The conceptual unity of a comprehensive theory – a theory which rests only on a few principles and which is as free as possible from the specific circumstances under which we perceive natural phenomena – makes the results of physical investigation independent of place and time, of the individuality of the investigator, and of nation and culture.[8] According to Planck, a unified world view is infinitely superior to the idea of an adjustment of our theories to the facts, as demanded by Mach[9], for physical universalism frees our cognition from the contingencies of human existence and leads to a constant reality behind the variable and manifold phenomena of the senses:

> "As I have tried to show, the constant unified world view is precisely the fixed goal which the real natural science, in all its transformations, constantly approaches ... This constant, independent of any human, or rather, of any intellectual individuality, is just what we call the real." (Planck, 1965, p. 49)

According to Duhem, the formal language of physics consists of mere symbols and leads away from reality; according to Mach, this language achieves the highest possible adjustment of our ideas to the facts only at the price of abstraction, simplification, schematization and idealization;[10] according to Planck, by contrast, it is this language that first leads to the cognition of a constant reality. Here there are obviously diametrically opposed conceptions of what is real at play.

For Duhem or Mach, reality lies in immediately observable phenomena. Duhem, the experimental physicist, identifies them with the results of observation gained in physical experiments; Mach, the phenomenalist, sees them in the end in original elements of our sensations – in sense-atoms, so to speak, instead of the physical atoms of Boltzmann, Planck, Einstein, Rutherford or Bohr. For Planck, by contrast, there is a constant reality to be found behind the changing play of phenomena of sense. This play of the phenomena of sense is conditioned, on the one hand, by the constitution of our faculty of cognition from which physics ought to emancipate itself as much as possible and, on the other hand, by the unchangeable laws of the underlying reality, the constitution of which only becomes visible through this emancipation.

The opposition thus sketched between Planck's scientific realism and Mach's empiricist or Duhem's instrumentalist position, however, appeared long before the turn of the 20th century. It is characteristic for the epistemological dispute surrounding the question of how modern physics relates to reality which has been carried on since the beginning, i.e., since Copernicus and Galileo. The Copernican world view, on behalf of which Galileo was tried, and Galileo's new theory of motion had to overcome

---

[8] (Planck 1965) p. 45.

[9] See Mach (1926), p. 164.

[10] See Mach (1926), p. 455.

Aristotelian physics and scholastic philosophy. Aristotelian physics was regarded as intuitive and close to experience, and it made no use of mathematical and technical methods. The geocentric picture of the world too was intuitive and close to experience, and it did not come into conflict with the Bible. Copernicus, Kepler and Galileo turned away from Aristotelianism in order to take up the Pythagorean and Platonic tradition. The applied mathematics practiced in the school of the Pythagoreans had always been regarded as an esoteric science. The innovation of Galileo's experimental method was that it allowed for an immense expansion of the Pythagorization of nature by tailoring natural events under laboratory conditions to the applicability of mathematics. In addition, Galileo introduced optical instruments such as the telescope into astronomy in order to render existing observational data more precise and conduct new kinds of observations. This, of course, called the objection of the Aristotelians to the scene that, rather than serving the observation of nature, devices such as the telescope and mechanical experiments served the generation of artificial phenomena through technical methods. This objection too is still found in the later debates about the realist or instrumentalist interpretation of physical theories. In this connection, I only want to recall Eddington's provocative question of whether the experiment perhaps stretches nature onto the Procrustean bed.[11]

Since the establishment of modern physics, the debate surrounding the realist interpretation of physical theories never subsided. Again and again, new epistemological objections were raised against metaphysical assumptions regarding physical reality. We can roughly distinguish five periods, in each of which this debate found a different expression – depending in part on the historical state of physics and in part on the philosophical and cultural environment. In this regard, however, the fundamental critique of a Pythagorean-Platonist understanding of reality must be strictly separated from objections against the metaphysical assumptions that go hand in hand with specific physical theories. Failing to do so and grounding at times the former and at times the latter in the requirement of intuitiveness is not exactly conducive to an epistemological clarification of the issue.

(a) What shaped modern physics at its outset was its confrontation with late-scholastic Aristotelianism. The conflict of the Copernican world view with the Bible led to theologically motivated attempts at defusing the truth-claim of the Copernican system and of Galileo's new astronomical findings through instrumentalist objections. Epistemological weaponry was used in order to defend ecclesiastical dogmas. Accordingly, Galileo's main works presented the arguments for the new world view in the cloak of artistic dialogues which appeal to the readers not to follow the Aristotelians in believing in appearances on the one hand and in tradition on the other, but to satisfy themselves through autonomous thought and logical analysis of the truth of

---

[11] See Eddington (1949), p. 106. As a more recent work which treats this question with the required discrimination, Hacking (1983) deserves special mention.

the new counter-intuitive discoveries concerning the structure of the world and the movements of bodies.

(b) The age of the Enlightenment actually carried out Galileo's demand for autonomous thought in the area of natural science. This age was shaped by the development of classical mechanics into an axiomatic theory, as first presented in Newton's *Principia*, and by epistemological disputes regarding the metaphysical foundations of the science of nature. The dominant philosophical currents of the 17th and 18th centuries were: the rationalist metaphysics which reigned from Descartes through Leibniz and Wolff until the pre-critical Kant; French materialism; the Cambridge Neo-Platonism which strongly influenced Newton; and British empiricism, with which, for a long time, Newton's metaphysics was falsely associated. In this period, the language of physics is largely dominated by metaphysical concepts such as "innate force", "substance" and "absolute space". These concepts were the subject matter of fierce metaphysical disputes, as, for example, in the debate between Leibniz and Clarke. Kant sought partly to criticize them and partly to integrate them into a metaphysics of experience which was tailored to suit the structure of Newtonian mechanics and which set entirely new standards for the intuitiveness of a scientific theory. This resulted in a shift of the debate away from the metaphysical issue of the constitution of material and immaterial reality and towards epistemological problems.

(c) In the 19th century, the debate surrounding Kant's theory first produced German idealism, then the Romantic philosophy of nature, and finally the strict delimitation of empirical natural science against all metaphysics on the part of empiricistically oriented physicists. Faraday founded field theory and thermodynamics. Maxwell's electrodynamics and the kinetic theory of gases were established. The language of physics was extended with the concepts of "energy" and "entropy". Maxwell and Boltzmann still based their theories on intuitive mechanical models. Maxwell developed a mechanical model of ether as the carrier of electromagnetic waves. Helmholtz sought to transform Kant's metaphysics of experience into an epistemology based on natural science. In this regard, he attempted, in particular, to explain the forms of intuition of space and time in terms of a physiology of the senses. Mach polemicized against the assumption of the existence of atoms and demanded in an appeal, for all people, to the atomist and scientific realist Newton that atomic theory later be replaced by a "more natural intuition" (Mach, 1883, p. 466). Ironically, the realism debate at the end of the 19th century had thus returned to the long-known instrumentalist objections against the understanding of reality of physical science.

(d) With the scientific revolutions at the beginning of the 20th century, physics itself took an epistemological turn critical of metaphysics. The physicists' view of the empirical possibilities of discovering the structure of reality changed drastically, and the realism debate received new tinder. Einstein subjected Newton's conceptions of space and time to a fundamental critique which led to the operational definition of simultaneity and to new concepts of space-time. Many Neo-Kantians regarded this critique as an attack against Kant's theory of intuition. Carnap and Reichenbach turned away from Neo-Kantianism and, following Mach's footsteps, established logical empiricism. Finally, in the course of the rise of National Socialism, Einstein's theories were reviled as an impertinence to sound common sense. (See Hentschel, 1990, p. 122.) In addition, the experimental findings of atomic physics forced the abandonment of classical radiation theory and of the classical idea of a physical object as incompatible with the property structure of quantum mechanics. Einstein in turn found Heisenberg's operational interpretation of subatomic processes, Bohr's concept of complementarity and the Copenhagen interpretation of quantum mechanics unacceptable. He was never able to submit to Bohr's and Heisenberg's view that quantum mechanics forces us completely to redefine the relation of language to physical reality.[12] Einstein held fast to a metaphysical understanding of physical reality which is rooted in traditional modern metaphysics and which Bohr and Heisenberg regarded as obsolete because it was tailored to suit the structure of classical physics.

(e) The circle is closed with today's postmodern tendency to declare the cognitive ideals of modern physics to be obsolete and to interpret the development of physics in a historicist way. For the postmodern sociologist and historian of science, physics is tailored to deliver a necessarily non-unified set of instruments of technologically applicable models; physical theories and their experimental foundations are mere constructs of human thought and action; and our physical world view is to be regarded as socially conditioned and culturally relative. This view in the end makes virtue of a necessity in that it turns the *lack* of understanding into an *unwillingness* to understand. Without entering into the details regarding the recent constructivist misunderstandings of physical theories,[13] I only want to note one thing. They are supported by an anti-Enlightenment frame of mind, the disastrous effects of which ought to be recalled especially in Germany. The fact that some constructivists present this view with the conviction of thereby continuing the work of the Enlightenment does not improve the situation.

Nevertheless, the most recent critique of realism raises a valid point. Recalling Planck's 1908 plea on behalf of scientific realism, one must admit that with the appearance of relativity and quantum theories, the unified and constant reference of physics came up for debate for new and structural reasons. Quantum processes are in principle incompatible with the locality assumptions of classical physics. They cannot be explained in terms of the relativist condition of Einstein causality, according to which signals cannot be transmitted at a rate faster than the speed of light. Until the present day,

---

[12] See especially Bohr's Como lecture (Bohr 1928) and his recapitulation of the discussions with Einstein (Bohr 1949); Heisenberg (1942; especially Section 1) and Heisenberg (1960).

[13] On this issue, see Scheibe (1997).

there exists neither a completely convincing quantum theory of measurement nor a satisfying approach to the unification of quantum theory and gravitational theory. For this reason, physical science is today indeed fragmented, and one must ask seriously whether this perhaps indicates limits to the theoretical unification of physics which threaten the *universalism* demanded by Planck.[14]

It is true that modern physics rests on a unified symbolic *language* of mathematical functions and concepts of physical quantity. The existence of this formal language, however, in no way ensures that all physical descriptions of systems can also be embedded into a *unified axiomatic theory*. The scale of a physical quantity – for example, the length or mass scale – always covers the domains of several theories, the axiomatic foundations of which are mutually incompatible. It covers the dimensions of quarks and atomic nuclei as well as macroscopic bodies and even the estimated mass and extension of the universe as a whole. Today's physics is far from being able to capture all these objects by means of a unified theory. Faced with the lost unity, however, physics is even further away from being able completely to dispense with the traditional epistemic claims of Copernicus, Galileo or Newton. Instead, one tries to embed the manifold theories and models of today's physics into a *unified informal language*. This language is flexible enough to iron out, to a certain extent, the formal breaks in the physical description of reality at the level of semantics – in the belief that a future axiomatic unity of physics will arise, the possibility and desirability of which functions as a regulative principle of theory formation. The informal language of physics is still *quasi-classical* and thus in relation to quantum processes or relativistic processes sometimes misleading. It induces one to talk about quantum objects or about the universe as one talks about objects of daily life. It comprises such intuitive expressions as "wave" and "particle" and applies them to subatomic events *without*, however, attaching to this use a claim to a complete classical description of an object. In addition, this language describes the evolution of the universe beginning with a Big Bang as if one were dealing with a physical system observable under laboratory conditions. Before dealing with its manner of function and its capabilities, however, we shall present, with recourse to Kant, *what* precisely was deplored as a loss of intuitiveness in the transition to the physics of the 20th century.

---

[14] Nancy Cartwright's works, in particular her book *How the Laws of Physics Lie* (Cartwright, 1983), provide tinder for this question. An additional characteristic feature of the structure of today's "postmodern" physics is the complexity of the phenomena in question. The new science of complexity required by this fact is described in Chap. 10.

2   Functions of Intuition

The founders of contemporary physics generally referred to Kant when they termed classical physics intuitive and non-classical physics non-intuitive. It was not only the engagement with Mach's critique of metaphysics which left its mark on the epistemological attitudes of Planck, Einstein, Bohr, Heisenberg and Pauli,[15] but also the debate surrounding Kant's theory of nature initiated by the Neo-Kantians. What were the physicists able to find in this theory of nature regarding the intuitiveness of physical objects, or what semantic functions does Kant ascribe to intuition in the interpretation of the formal language of physics? This question is decisive, for *prior* to Kant's critical philosophy there existed *no* unified theory of intuition.[16] *After* the reception of Kant's *Critique of Pure Reason*, by contrast, essentially *every* educated person, whether a Kantian or not, took "intuition" more or less to refer to the pure forms of intuition, space and time, described by Kant in his *Transcendental Aesthetics*.[17] The latter are cognitive faculties a priori which are among the conditions of the possibility of experience. At this point, three fundamental remarks about Kant's theory of intuition are required.

(a) Kant's identification of space and time with forms of intuition, which are at the same time subjective conditions of the possibility of objective experiences, rests on certain basic assumptions of Newtonian physics. Yet it emerged from an engagement of many years with the metaphysical debates of the 18th century about space and time, in particular with the Leibniz-Clarke debate.[18] With his theory of intuition, Kant wanted to preserve absolute space as an ideal frame of reference for the construction of inertial motions

---

[15] See Einstein 1949, Planck's lecture, *Die Einheit des physikalischen Weltbilds* (Planck, 1965), and Bohr (1923). See also Falkenburg (1998a). Pauli profited from Mach's influence especially in the years 1923/24, when the old quantum theory fell into a crisis because of confusing findings from atomic spectroscopy which were due to the genuinely non-classical degree of freedom of "spin". His godfather was Ernst Mach, and Pauli expressed the following about his relationship to Mach: "He evidently was a stronger personality than the catholic priest, and the result seems to be that in this way I am baptized 'antimetaphysical' instead of catholic." (Pauli, 1994, p. 13)

[16] The rationalist doctrine of ideas from Descartes to Leibniz and to Wolff did not make such a sharp distinction between two kinds of representations as Kant did since his *Dissertation* of 1770, i.e., between discursive representations or concepts and intuitive representations or intuitions. Neither was intuition prior to Kant necessarily associated with our cognitive faculty of representing particular objects in space and time.

[17] In particular, the founders of modern logic and mathematics also referred to Kant, whether critically (Cantor and Frege) or affirmatively (Hilbert). The Marburg school of Neo-Kantianism (Cohen and Natorp) for its part eliminated Kant's theory of intuition from his philosophy of science.

[18] See, for example, the introductory chapter in Friedman (1992) or Beck (1969), Chap. XVII, p. 438.

without ignoring Leibniz's critique of the concept of a *real* absolute space. He wanted to reconcile absolute space with Leibniz's relational conception of space and time, according to which space and time are mere relations of material phenomena – the orders of the coexistence and succession of appearances of nature.

(b) Kant was originally a metaphysical realist whose striving for unification went far beyond a scientific realism in the Planckean sense discussed earlier. His pre-critical cosmology was supposed to integrate the diverging metaphysical views of the 18th century about space, time and matter according to the best reasons and counter-reasons into a unified metaphysics. This attempt eventually led to such insuperable intra-theoretical difficulties that the only way out for him was to develop, instead of an *objective* theory of space and time as real things, properties or relations in nature, a *subjective* theory of space and time as mere forms of our sensibility.[19] Expressed in Planck's terminology of 1908, Kant had now arrived at the conception that space and time are *anthropomorphic representations* – representations, however, of which we cannot free ourselves in the course of physical theory formation, according to Kant, because they are conditions of the possibility of *all* objective experience.

(c) As the subjective form of intuition which is given a priori in advance of all experience, space determines the structure of experience throughout with a *Euclidean metric*. Kant was aware of the (purely logical) possibility of a non-Euclidean geometry, but he rejected it for the purposes of the application of mathematics in a physical cosmology which is supposed to deal with *objects of experience*. According to Kant, the models of a non-Euclidean geometry are not constructible in intuition, and this is why he does not grant them the status of a "real" possibility.[20] In this regard, his theory of intuition is, from a contemporary view, beyond repair – even if one interprets it in such a way that the doctrine of space and time of the *Transcendental Aesthetics* is only valid for a common experience from which scientific experience could be quite distinct. However, the experience, the structure of which is analyzed in the *Critique of Pure Reason*, is *not* common experience, but scientific or systematic experience, the experience that forms the basis of empirical natural science.[21]

Because Kant's theory of intuition refers to scientific experience, it unavoidably comes into conflict with the special and general theories of relativity. Carnap and Reichenbach were only consistent when they turned away

---

[19] On this issue, see Falkenburg (2000a), Chap. 3.

[20] On this issue, especially Friedman (1992), p. 92.

[21] This is emphasized especially by Friedman (1992), who is following Neo-Kantianism in this regard. See also Kant's comparison between metaphysics with the course of a science, which he wants to support with his theory of experience, and the systematic procedure in empirical natural science (Kant 1787) p. B XIII, as well as his definition of science as cognition with systematic unity (Kant 1781/1787) p. A 860/B 832.

from Neo-Kantianism in order to establish a scientific empiricism. They held the view that Einstein's theories of relativity indicate that the structure of space-time is *empirical*. From their perspective, Kant's theory of intuition in particular had the function of grounding the axioms of Euclidean geometry – for example, the synthetic-aprioric character of the parallel axiom.[22] Einstein's theories, by contrast, demonstrate that there exist physical alternatives to Euclidean geometry about which one can make an empirical judgment based on measurements. Thus, even if they are a priori, Kant's forms of intuition are not conditions of the possibility of *all* experience, that is, they are not valid with unconditioned necessity. From the perspective of logical empiricism, the structure of space-time is to a large extent empirically determined (aside from certain elements of free choice, such as the Einstein convention regarding the synchronization of clocks[23]). The Kantian a priori is therefore at least relativized,[24] if not completely eliminated. According to Carnap, mathematical geometry, which rests on analytic judgments and the axioms of which are determined a priori, must be distinguished from physical geometry, which is synthetic and the choice of which rests on experience, as follows:

> "Mathematical geometry is a priori. Physical geometry is synthetic. No geometry is both. Indeed, if empiricism is accepted, there is no knowledge of any sort that is both a priori and synthetic. ... A clear distinction here is essential if confusion is to be avoided and if the revolutionary advances in the theory of relativity are to be understood." (Carnap, 1966, p. 183.)

From the perspective of the disloyal Neo-Kantians, not much more could remain of Kant's theory of intuition than a *psychological* conception of space and time which would have to be strictly distinguished from a mathematical or physical theory of space-time. The reproach of psychologism, first leveled against Kant by Hegel and Fries and later picked up by the Neo-Kantians of

---

[22] See, for example, Reichenbach, (1951) p. 40. The theory of intuition is, after all, an essential component of Kant's doctrine of synthetic judgments a priori which refer to objects of experience (and are thus, according to Kant, synthetic) and yet have apodictic necessity (because they are a priori, i.e., because they are established prior to all experience). Hence it follows, for example, from the intuitive evidence of the representation of two parallel straight lines that the parallel axiom is a synthetic judgment a priori and thus necessarily true.

[23] In his work of 1905, Einstein must determine *by definition* that the time necessary for light to travel from A to B is equal to the time that it takes to travel from B to A (Einstein, 1905; §1). Beginning with Reichenbach, this definition spawned a debate about conventionalism among philosophers of science. Regarding the consequences of a non-standard definition that deviates from the Einstein convention, see Friedman (1983), p. 165.

[24] This was at first Reichenbach's position; see Reichenbach (1920).

the Marburg school,[25] seems to suggest itself. For the forms of intuition of space and time are for Kant nothing but subjective cognitive faculties, which have the task of bringing everything we perceive into the orders of coexistence and succession. It is true that, according to today's knowledge, our perceptual space is not exactly Euclidean, but it does have the structure of a three dimensional topological manifold, and gestalt-psychological experiments show that in our culture we are accustomed to assume its Euclidean character. Accordingly, even today, it is often claimed that the non-intuitiveness of a relativist space-time is a purely psychological problem which can be remedied to a certain extent by the use of two dimensional diagrams which make Minkowski space or a curved space-time intuitive. From this perspective, the deanthropomorphization of physical concepts worked out by Planck in his 1908 lecture has advanced further than Kant ever thought possible. Yet it is doubtful that the only function of intuition in physical theory is to make space-time structures of higher dimensions intuitive for didactical purposes by means of diagrams. Already in Kant's theory, the function of intuition cannot be reduced to the task of proving that the axioms of Euclidean geometry are synthetic judgments a priori. For this reason, it is in no way settled that with the disappearance of this task only the psychological function of making things intuitive remains.

In order to characterize the functions of intuition in Kant's theory of nature, one must go back to the distinction between *intuitions* and *concepts* which is fundamental to Kant's theory of space and time as subjective forms of intuition. With this distinction, Kant made the decisive break with the rationalist doctrine of ideas which in the tradition from Descartes to Leibniz, Wolff and Wolffianism had formed the common basis of every epistemology. The fundamental assumptions of this doctrine of ideas were even found in British empiricism. The break with it was nothing else but a scientific revolution in Kuhn's sense. After all, Kant himself spoke in the preface to the 2nd edition of the *Critique of Pure Reason* of a "revolution of thought". This revolution brought about the end of the school of the Leibniz-Wollfian philosophy and the emergence of German idealism. From 1770 on, Kant considered intuitive representations, which have immediate evidence for us, and discursive representations, which we connect step by step in logical operations, no longer, as Descartes, Leibniz or Wolff did, as fundamentally related kinds of representations which differ at most in the degree of clarity and certainty of their individual marks. He now regarded them as two kinds of representations

---

[25] Thus Cohen (1902; p. 23) as well as in earlier writings distances himself decisively from Kant's theory of intuition. According to Cohen, Kant's theory of intuition is merely a psychological theory which must be replaced by a purely logicist grounding of the exact sciences. See also Falkenburg (2000a), p. 309. Since Russell, the reproach of psychologism against Kant extends beyond the theory of intuition to include the traditional logic of concept and judgment as well. The reproach is still found today, for example, in the by Tugendhat and Wolf (1986).

distinct in principle which derive from the understanding and from sensibility as two distinct cognitive faculties and which possess the following opposite formal marks: concepts are general representations; they are composed of at most a finite number of general marks (or predicates) and subsume their partial representations *under* them. Intuitions, by contrast, are individual representations; they are given individually in space or in time as pure forms of sensibility; they represent infinite given quantities (with the characteristics of the mathematical continuum) and can comprehend an arbitrary number of partial representations *within* them.[26]

In other words: concepts are *abstract* and *symbolic* and *subsume* specific contents or objects under them; intuitions, by contrast, are *concrete* and *image-like* and *represent* specific contents in them. According to Kant, the cognitive achievement of intuition primarily consists of the fact that it is our *individuating faculty*: it allows us to distinguish individual things in space and time. The insuperable difficulties which Kant's pre-critical cosmology encountered indeed resulted primarily from a problem of individuation which Kant was only able to solve with his theory of intuition of 1770 and which I shall sketch later. According to the critical Kant, only intuition supplies concrete conceptual contents in space and time and hence concrete objects for abstract concepts of the understanding. It supplies the *domain of application* for abstract concepts which can have objective reality, and thus intuition plays a central role in the epistemology of the *Critique of Pure Reason* for the constitution of objects of experience.[27] Accordingly, intuition is constitutive for the production of the objects of any scientific theory, be it mathematics, physics or the metaphysics with the course of a science Kant wanted to establish with his critique of reason. In his theories of mathematics and nature, intuition has the semantic function of supplying the objective realms for the abstract and symbolic concepts of formal theories. Since, for this purpose, he did not yet have abstract set theory at his disposal, he took recourse to space and time. Their identification with pure forms of intuition was supposed to make possible the interpretation of formal theories in the following way: the objects of a formal theory can in each case be formally constructed as finite limitations of the infinite space of intuition and generated in thought successively in time. Kant's theory of intuition was thus supposed to fulfill three distinct semantic functions:

(a) It served structurally to extend his logic, which is based on the traditional doctrine of judgment and is relatively wanting in structure.[28] This

---

[26] See Kant (1770), §14 and §15, (Kant 1910, Vol. II, p. 398-406); and in parallel with this, Kant (1781/1787) p. A 22/B 37, A 31/B 46.

[27] Kant's distinction between "logical" and "real" possibility is of central importance in this connection; Plaass (1965), p. 55, Friedman(1992) p. 94-104.

[28] Yet it is not confined to the limited structure of monadic logic: Kant was, of course, familiar with relations, and he was also able to express them by means of the traditional forms of judgment. Friedman (1992, p. 63, by contrast, claims: "For Kant logic is of course syllogistic logic or (a fragment of) what we call

extension made it possible for him, among other things, (i) to define potentially infinite domains of individuals, (ii) to quantify over them, (iii) to generate continuous structures and with all of this (iv) to ground arithmetic and Euclidean geometry. At first, this concerns a *formal-semantic* extension which does not yet refer to objects of the empirical sciences. In principle, it is supposed to achieve a similar result as set theory does in contemporary mathematics. As the discussion of the foundations of mathematics in the 20th century has shown, this result can likewise not be achieved on the basis of a symbolic logic alone.[29]

(b) In identifying the pure forms of intuition with space and time, Kant solves a problem which occupied him in his first work, that is, the problem of providing a metaphysical reason for the fact that mathematical physics can be applied so successfully to appearances of nature. The *Critique of Pure Reason* offers the explanation that the objects of mathematical physics must always already feature the spatio-temporal structure of the objects of our cognition because they are constructed in pure intuition. The price for this solution, however, is that for Kant there only exists *applied mathematics* which is limited to *potentially infinite models of arithmetic and of geometry*. For the critical Kant, actually infinite sequences or non-Euclidean spaces are *logically* conceivable, but their logical concepts are abstract and without sense and reference.

(c) The pure forms of intuition are supposed to make possible the *individuation* of objects *in* space and time. For decades, Kant had pondered the question of how the existence of things of the same kind, which are the object of natural science, such as crystal formations or identically shaped lumps of gold can be reconciled with the relational theory of space and time for which Leibniz had named good reasons. The most striking example of objects of the same kind, the distinguishability of which requires explanation were for Kant so-called *incongruent counterparts*, which like one right and one left hand feature an opposite sense of helix and which can be transformed into one another only by means of a spatial reflection.[30] According to Kant, the distinction between such forms is not a relational property which can be defined in a combinatorial way,[31] but an *absolute property* which is fixed relative to

---

monadic logic." The view, prevalent in the Anglo-Saxon literature, that traditional logic is limited to monadic predication, goes back to Russell's critique of Leibniz. Compare, by contrast, Ishiguro's (1990) reconstruction of Leibniz's relational logic as well as Strawson's (1952), Chap. 6) remarks regarding the translatibility of traditional syllogistics and the square of oppositions into modern predicate logic.

[29] For this issue, see Hallett (1995).

[30] Kant (1768).

[31] A purely combinatorial interpretation of the right-left asymmetry was suggested by Weyl (1949, Chap. 14) in connection with Leibniz's project of a geometric characteristic. Kant's work of 1768 criticizes the project of an *analysis situs* (analysis of the relations of location) on which the geometric characteristic was

a presupposed oriented space of reference. His theory of space and time as subjective forms of intuition was supposed to solve this problem as well; in fact, it can be shown that he *developed* it, *in order* to solve this problem.[32]

It is in particular the second semantic function, i.e., the desired solution of the problem of application of mathematical physics, which brings Kant's theory of intuition into conflict with the special and general theories of relativity. Binding mathematics to a cosmology intrinsically tailored to the Euclidean structure of the space of intuition considerably limits the possibilities of mathematical model building and with it the structural extension of logic by means of a formal semantics. To a certain extent, the question of why mathematical physics is applicable to appearances of nature could be answered by means of an analysis of the structural characteristics of Galileo's experimental method rather than by means of a theory of space and time which is a central constituent part of an epistemology. Had Kant recognized this, he would not have had to restrict unnecessarily his theory of mathematics in comparison to Leibniz. The individuating function of intuition, on the other hand, is *prima facie* independent of Kant's problematic attempt to ground geometry and arithmetic in such a way that the concepts of mathematical physics are a priori applicable. This function, however, refers to three quite distinct structural characteristics of the domain of individuals of an empirical natural science. For, according to (3.), intuition guarantees the ability to distinguish:

(i) individual objects in space and time which can be individuated on the basis of their spatiotemporal development and which are considered as concrete carriers of partly changing and partly permanent properties;[33]

(ii) objects in nature which are of the same (or *nearly* the same) kind which may be regarded as representatives of natural kinds and made the basis of the formation of empirical classes – mathematical concepts are obviously applicable to them already *without* any experiment; and

(iii) mirror-opposite formations in nature, the internal distinction of which is indeed a tough nut to crack for any relational theory of physical properties – this distinction could be much better explained in an 'essentialist manner' within the framework of a metaphysical realism, i.e., either through

---

to be based. The argument of incongruent counterparts is supposed to show that Leibniz's primitive relations of equality (equisidedness) and similarity (equiangularity) are insufficient to ground (physical) geometry.

[32] See Falkenburg (2000a, Chap. 3). This issue is also worked out from a purely mathematical perspective by Buroker (1981, 1991): (the other contributions to this volume deal with the left-right problem from the perspective of the philosophy of science, but do not enter into the role it plays in Kant's development.)

[33] According to the *Critique of Pure Reason*, however, intuition alone is not sufficient for this purpose; one also needs schematized categories and the principles of pure understanding.

the assumption of intrinsic physical properties or through the assumption of an absolute space in Newton's sense.

From Kant's perspective, the first two problems of individuation are very closely connected, since it follows from Leibniz's relational conception of space and time and Leibniz's principle of indiscernibles that homogeneous things in nature are not only indistinguishable but identical – if one does not ascribe to them with Leibniz some hidden inner differences and thus declare their homogeneity to be only apparent.[34] For a long time, Kant thought it possible to solve this problem by means of a relational theory of the world as a whole. In 1768, however, he discovered the third problem of individuation which blocked this avenue.

It is difficult *not* to regard these three respects in which individual objects in space and time can differ as *empirical* properties of material things, similar to the way in which Carnap and Reichenbach regarded the metrical structure of physical space-time as empirical. Contemporary subatomic physics has shown that the *characteristics of individuation* of constituent parts of matter can also be completely different than Kant thought of them following his theory of intuition. Quantum mechanics and the recent quantum field theories of elementary particles require a deep revision of the classical view of the distinguishability of the constituent parts of matter which Kant takes as his basis. According to Heisenberg's indeterminacy relation for position and momentum, the spatiotemporal properties of subatomic particles are determined only probabilistically. Quantum systems cannot be individuated on the basis of the spatiotemporal development of their states. Subatomic particles of the same type are indistinguishable and are subject to a non-classical statistics (Fermi or Bose statistics). The difference between certain types of radioactive decays and their mirror image in turn is bound up with intrinsic particle properties such as parity which have an effect on the observable decay rates, and which may form superpositions. If one wants to assume with Kant that individuals can be individuated *a priori* through intuition, one must ask oneself, how it is possible that in the field of quantum theory, this assumption must obviously be *surrendered* for empirical reasons.

And yet, it is precisely due to this failure that intuition still has an unrenounceable semantic function for contemporary physics. Quantum systems are just as non-intuitive as a relativist space-time inasmuch as we cannot represent them to ourselves at a given time in three-dimensional space. In the face of relativist physics and quantum physics, however, intuition fails in completely different ways. It is true that cosmological models with curved space-time conflict with the global structure of the three-dimensional space of intuition, but at least locally, they have the structure of objects of our intuition. Descriptions of quantum systems, by contrast, are incompatible with

---

[34] According to Kant's *Nova dilucidatio*, of 1755, however, this would amount to seeking fault in a truism ("hoc enim esset nodos in scripio quaerere") (Kant, 1910, Vol. I, p. 410).

the local structure of the objects of our intuition. Quantum systems cannot be represented in the space of intuition, because they cannot be individuated, or because they are lacking the usual properties of locality and separability of the objects of our intuition. According to the predictions of a quantum theory, a subatomic particle which has been localized by a position measurement is after a short period of time forever in a non-localized state; and to this day, it cannot be fully explained why a quantum system, by way of measurement, reveals a well-defined physical property, if prior it did not already have a well-defined value of the measured quantity.[35] Thus, in the end, we need intuition in order to relate the abstract models of a quantum *theory* to experimentally generated quantum *phenomena* in a physics laboratory. To put it more precisely: in order to apply the formalism of a quantum theory to experimental results, one must be able to connect the abstract system descriptions gained from this formalism with concrete models of individual quantum systems which also have an intuitive meaning.[36] With reference to Kant, Niels Bohr in particular emphasized this again and again.

3   The Graspability of Cognition

Bohr called the phenomena of subatomic physics which led to the development of quantum physics *individual*. With this he expressed the fact that by means of the experimental method, quantum phenomena cannot be analyzed further – for example, into the macroscopic equipment, by means of which a quantum phenomenon is measured, and a quantum *object* quite distinct from this.[37] Only the quantum *phenomena* observed in an experiment have the character of individual things or processes in time and space. For this reason, Bohr (and with him Heisenberg) took the view that the language in which the formalism of a quantum theory is interpreted must always fall back on concepts of *classical physics*. He regarded this language as *intuitive* in the sense of Kant's epistemology, as that language which allows for the description of the objects of scientific experience as individual objects in time and space and at the same time for the expression of their causal effects on one

---

[35] The most advanced attempt at a solution of the quantum-mechanical measurement problem arrives at the result that the coherence of superpositions is quickly destroyed during measurement. This gives the possible results of measurement a *classical* statistical structure. Yet it remains unexplained why during individual measurement one of them is distinguished and realized over the others. On the decoherence approach, see Giulini et al. (1996).

[36] In general, such models stand under quasi-classical constraints. There exists, for example, a chain of quantum models of atomic and subatomic structure which mutually stand in relations of approximation and which, in the end, stand in correspondence to Rutherford's classical description of the scattering of charged particles at the atomic nucleus. See Falkenburg (1995, Chap. 4) as well as Falkenburg (1998b,2000b).

[37] For this issue, see Falkenburg (1998a).

another. By contrast, Bohr again and again designated the formal quantum mechanics of 1925-26 which was first given an axiomatic form by von Neumann as *abstract* and *symbolic* (similar to how Schrödinger designated the reduction of the quantum-mechanical wave function during measurement). Following Kant, this would mean: taken just by itself, quantum theory is a formal theory without sense and reference, one that has no objective reality or *no concrete objects of reference*. Even following Duhem, one would understand the phrase, "abstract and symbolic", in this sense. There exists, however, a decisive difference between Bohr's and Duhem's views of physics. According to Duhem, *all* physical theories and quantities have merely an abstract and symbolic meaning. According to Bohr, by contrast, this is true *only* of axiomatic quantum theory and of quantum concepts or models without classical correspondence. Thus, Bohr is one source of today's common sense view, according to which only the physics after the turn to the 20th century is abstract and non-intuitive.

Bohr considered *classical* physics and its language as the decisive semantic frame of reference needed in order to make a quantum theory *concrete*. Thus, unlike Duhem, he represented a scientific realism at least with reference to the language of classical physics, one which may also be understood to a certain degree as an empirical realism in a Kantian sense. From Bohr's point of view, atomic reality can only be captured "complementary" means of description which are mutually exclusive, such as the spatio-temporal and the causal description of quantum processes. Evidently, this limited conception of the reality of quantum phenomena still draws on an understanding of intuitiveness which is decisively shaped by Kant.[38]

In two respects, Bohr's empirical realism was weaker than Planck's or Einstein's metaphysical realism: (a) It did not refer to the objects of a quantum theory, but only to the *language* in which quantum *phenomena* are expressed. According to Bohr, the usual concept of an object in physics loses its applicability in the subatomic realm, and only the intersubjective communicability of the experimental results can take its place.[39] (b) In addition, Bohr was fundamentally convinced that the break between the classical and the quantum theoretical description of reality signified an insuperable limit for the theoretical unification of physics – and that there are reasons in principle for this limit. For according to his view, Heisenberg's relations of indeterminacy indicate that the experimental method encounters natural limits in the subatomic realm.

If this view is justified, then these natural limits of the experimental method bring not only the unification of physics to a halt, but also the de-anthropomorphization of the physical concepts and with it Planck's metaphysical realism. But how are we supposed to interpret the concepts of physical theories in domains in which we can no longer employ intuitive quasi-

---

[38] For this issue, see Faye (1991).
[39] See Bohr (1948) and Chevalley (1994, 1995).

classical concepts? This leads to a hermeneutic problem closely connected to the question of how tenable realistic positions are at all.[40]

Today, we know that the interpretation of the quantum theoretical description of empirical reality in terms of complementary classical concepts does not reach nearly as far as Bohr's complementarity claims. Yet this did *not* lead to the result that descriptions of quantum systems without any classical correspondence count as abstract and symbolic models without sense and reference. Today's physics is able to make even genuinely non-classical concepts such as spin and typical quantum phenomena such as the double slit experiment or the Einstein-Podolsky-Rosen correlations intuitive through a flexible use of informal concepts which reveal their origin from classical physics. The informal terminology of current particle physics follows the concept of a particle of classical physics step by step – even though it is clear to every physicist that this concept cannot cope with the referential objects of a quantum field theory. One speaks of a "particle" and means the energy or charge quanta of a quantized field which are localized by means of a particle detector. One speaks of a "particle track" and means the results of repeated position measurements. One speaks of "virtual particles" and means contributions to the perturbational expansion of an abstract scattering amplitude which in principle cannot be isolated experimentally.

In general, physicists like to devise a pictorial language when dealing with abstract models, one which partly serves to popularize the theory but which also furthers the emergence of crude misunderstandings. In quantum chromodynamics, by means of which the protons or neutrons in the atom can be described as bound systems, one speaks of "confined quarks" and "asymptotic freedom"; by this one means to refer to a binding energy which increases with growing distance and disappears at extremely small distances. The talk of "superstrings" appeals to our idea of vibrating strings. General-relativistic cosmology also employs pictorial expressions such as in the talk of the *Big Bang* or in the claim that within a black hole space and time *switch roles*. The abstract and symbolic contents of today's physical theories cannot be adequately represented in intuition or expressed in quasi-classical language. For all this, however, one does not dispute their concrete sense and concrete reference. At least indirectly they are connected to observable phenomena. The experimental experience on which they rest is theory-laden, but it is still experience. The pictorial language of physicists is not intended to be taken literally; it is merely designed to make the non-intuitive and experientially distant content of today's physical theories and the corresponding experimental results *graspable*. As I later want to show by means of the example of the Feynman diagrams of a quantum field theory, this making graspable does not primarily serve the popularization of non-intuitive theories and models of relativistic cosmology or quantum physics for non-physicists. It primarily

---

[40] I discuss this problem in Falkenburg (1993), with reference to a case study of the spatial structure of subatomic centers of scattering.

serves to make complicated formal methods *more manageable* for physicists themselves.

At the basis of the pictorial language of physicists, there is obviously an understanding of intuitiveness that is *substantially weakened* compared to Kant's or Bohr's requirement that the objects of our scientific cognition be representable in intuition. In conclusion, I would like to make a first attempt at specifying this weakened understanding of intuitiveness by returning once more to Kant's epistemology. In the introduction to his lectures on logic, Kant always presented a *doctrine of the perfections in cognition* by means of which he completed the doctrine of cognition of the Leibniz-Wolffian school according to his own epistemological principles. This doctrine comprised a canon of cognitive ideals which Kant regarded in part as constitutive for cognition in general and in part as regulative principles of the expansion of cognition.[41] Kant systematized this canon in accordance with his table of categories into (a) universality, (b) distinctness, (c) truth and (d) certainty of cognition. What was new in this was that besides the "logical" ideals of cognition of the rationalist tradition, the system also included "aesthetic" ideals, the treatment of which was based on Kant's theory of intuition, a theory which, of course, had marked the break with the rationalist doctrine of ideas. The logical ideals of cognition concerned what today are called logical completeness and conditions of adequacy of theories. The aesthetic ideals of cognition, on the other hand, concerned our present topic, that is, the *subjective graspability of our theoretical cognition*.

Kant understood the "aesthetic universality" of concepts, judgments or theories as the applicability in paradigmatic cases which are generally accessible and can serve to popularize scientific theories. "Aesthetic distinctness" for him referred to the existence of examples *in concreto* which are given as concrete representations in time and space or intuition. He understood "aesthetic truth" as the mere plausibility which in certain cases can also be deceptive, while he took "aesthetic certainty" to refer to the certainty of sense perception. According to Kant, these four aesthetic ideals of cognition are usually in conflict with the logical ideals. Cognition, which is ideal in *every* respect, does not exist in his view. The logical ideals of cognition require that a theory describes its objects in a logically complete and adequate way. The aesthetic ideals of cognition, by contrast, require that the concepts and claims of a theory refer at least partially to something that is familiar, intuitive and graspable through sense perceptions. Kant himself knew that both are seldom obtainable at the same time. In particular, he knew that his own main work, the *Critique of Pure Reason*, was far from being able to satisfy the claim to "aesthetic perfection". This is why he wrote the *Prolegomena*, in which he moderated his demands for logical completeness for the sake of popularity.

---

[41] For the following, compare the Jäsche logic in: Kant (1910, Vol. IX), especially p. 36, and parallel passages in the Vorlesungsnachschriften (Kant, 1910, Vol. XXIV).

For the purpose of presenting the contents of a theory in a graspable way, a good popularization will attempt to sacrifice as little of the logical ideals of cognition as possible, least of all the ideal of *truth*. It would be an instructive exercise to check the research reports, textbooks and popular writings of physicist against Kant's canon of logical and aesthetic criteria of cognition! I cannot begin to do this here. Neither do I want to set up a typology of good popularizations based on Kant's distinctions. I would like to indicate, briefly at least, how Kant's aesthetic ideals of cognition can be related to the functions and limits of intuitive and visualizable concepts in contemporary physics.

The non-intuitive theoretical attempts and models of the theory of relativity and of quantum theory violate the condition of aesthetic distinctness. Neither a quantum system nor a relativistic cosmological model can be represented *in concreto* in intuition without certain sacrifices. The three-dimensional Euclidean models, by means of which one can visualize the structure of a non-Euclidean geometry, are still the most effective. They are intuitive and demonstrate moreover the consistency of the relevant geometry by means of a model in *another* geometry of lower dimensions. Thus they satisfy – in contrast to most popularizations – the conditions of aesthetic distinctness and truth. Popular examples such as the double slit experiment of quantum theory or the twin paradox of special relativity, on the other hand, popularize the so-called paradoxical traits of both theories; hence, according to Kant, they ensure the aesthetic universality of cognitions which are non-intuitive or which *lack* aesthetic distinctness. The condition of aesthetic truth (= plausibility) may be found in the symmetries of contemporary physics, which are de facto violated to a large extent. The aesthetic certainty (= ascertainment of sense) in turn is fulfilled by the experimental confirmation of theoretical predictions or the refutation of alternative theoretical explanations.

On the other hand, the pictorial language of many physicists merely functions to provide a surrogate for the intuitive concepts and models of classical physics which *fail* in the areas of relativistic physics and quantum physics. Pictorial expressions have an eidetic function. They suggest intuitive objects, and, relative to the background of classical physical ideas, they evoke the semblance of truth, i.e., they feign the reference to concrete objects in space and time. In the best case, this use of language is not only plausible because it connects to familiar ideas, but moreover it even harmonizes with the conditions under which formal descriptions of systems can be applied *in concreto* to experimental results. In the worst case, on the other hand, this use is grossly false and evokes completely misleading associations. Often, adequate and useless plausibilities coexist, and it requires a precise science-theoretic analysis in order to separate them.

Thus, for example, the talk of a wave-particle dualism of quantum systems or of their models remains confusing as long as it is not specified what in the domain of a quantum theory is to be understood by a "wave" or a "particle". If one wants to understand these concepts in the context of quantum theory in a strictly *classical* way, one will be induced to the grossly false claim that a quantum object "is" a wave and a particle at the same time and hence characterized contradictorily. If, on the other hand, one interprets certain models of the abstract formalism of a quantum theory by means of the physical quantities of "momentum" and "wavelength", one will be led to the de Broglie relation $p = \hbar k$, i.e., to a precise formal proposition with a well-defined operational content. In their works, Bohr and Heisenberg referred to the wave or particle *picture*. By this they meant two things: *on the one hand* the intuitive representation of concrete quantum phenomena which can be generated under various experimental conditions from one and the same physical system such as an electron beam – i.e., diffraction images or particle tracks; and *on the other hand* the quasi-classical modeling of these phenomena by means of classical measurement laws.

Now, if today, in the informal language usage of subatomic physics, there were hardly any more talk of waves but almost exclusively of particles, one thereby would not mean to refer to classically structured constituent parts of matter. Nevertheless, this manner of speaking does still establish a connection to central characteristics of the classical concept of a particle. The models of quantum mechanics or quantum field theory required for the description of many experimental phenomena of subatomic physics can be associated with *successor concepts* of the classical concept of a particle (see Falkenburg, 1995). For one can precisely state the experimental conditions under which subatomic scattering processes as well as the phenomena and measurement data resulting from such processes (particle tracks, scattering events, scattering cross-sections) have a classical analogue visualizable in a particle model. These experimental conditions are determined by relations of approximation between the quantum model and the classical model of a scattering process. Such relations of approximation not only provide intuitive models for subatomic scattering processes, but at the same time quasi-classical conditions of individuation for quantum models of scattering centers. In this respect, one can distinguish two core meanings of "particle" which by no means cover all the phenomena of subatomic physics, but nevertheless form the basis of continuing to hold on to an informal concept of a particle: (a) The informal concept of a particle refers to *constituent parts of matter which are localizable as discrete structures by means of scattering experiments*. The classical model for this is the Rutherford scattering of α-particles at the atomic nucleus. Subatomic structures can be described on the basis of scattering experiments by means of form factors which can be interpreted intuitively in a non-relativistic case and which even in a non-relativistic case are still valid as a measure of the resolution of scattering centers into discrete structures of a certain order of magnitude.[42] (b) The concept refers to the *cause of particle tracks* measured by means of particle detectors (bubble chamber, drift chambers). As the detailed analysis of the quantum description of a particle track shows, however, one thereby does *not* refer to a *per se existing and reidentifiable entity*, but rather only to the propagation of a quantized effect through repeated subatomic scattering processes which are subject to dynamic conservation laws and are demonstrated by means of position measurements in the particle detector. The propagation of this effect, however, is *reidentifiable* in a macroscopic environment by means of the observable particle track as if its cause *were* a particle in the classical sense.[43]

As an example of an excellent effort at making a theory plausible *without* reference to quasi-classical criteria of individuation, I would in conclusion like to mention the Feynman diagrams of quantum field theory. Quantum field theories are not only non-intuitive, they are also extremely difficult to handle formally. Feynman diagrams are intuitive graphs which have a precise symbolic meaning in the framework of the abstract formalism of a quantum field theory and which facilitate their handling immensely. Their intuitiveness, however, only *suggests* that they represent concrete processes in space and time. From an empirical point of view, they do not have any concrete physical meaning, for the portions of a scattering process which they symbolize cannot in principle be isolated by means of experiments. We are dealing with mere symbols with the formal function of instruments of calculation. Every Feynman diagram stands for a formal contribution to the perturbational expansion of a transition amplitude in quantum field theory, the square of the value of which provides the probability of a scattering process of real particles. The perturbational series as a whole is represented by an infinite sum of Feynman diagrams. In each diagram, the graphical representation suggests spatio-temporal events in which particles scatter against one another, are destroyed into vacuum and generated from vacuum. Each Feynman diagram of the perturbational series shows the same incoming and outgoing particles (they represent the scattered particles and the reaction products), but the intermediate states are distinct and become more and more complicated with the growing order of the perturbational expansion. Every particle in the diagram is represented by an open or closed line. Every line or loop of a Feynman diagram in turn can be translated, according to precise rules, into an algebraic expression which enters into the calculation of the perturbational series. In this manner, the calculation is facilitated immensely and, because of the intuitiveness of the symbolic representation, the procedure is at the same time readily grasped. Yet, whoever understands a Feynman diagram

---

[42] Ibid., Chap. 4, p. 140; the limits of the quasi-classical talk of constituent parts of matter in the area of a relativistic quantum field theory are worked out in Chap. 7, especially p. 284.

[43] Ibid., p. 253.

literally, i.e., interprets it as a representation of a concrete scattering process, goes astray.

References

Beck, L.W. (1969): *Early German Philosophy. Kant and his Predecessors* (Cambridge, Mass., Harvard University Press).

Bohr, N. (1923): Nobel lecture. English translation: "The structure of the atom", *Nature 112*: 29–44. Danish and English texts in *Collected Works*, ed. by Rosenfeld, L. and Rüdinger, E. (North Holland, Amsterdam 1972), Vol. 4, p. 425–482.

Bohr, N. (1928): "The quantum postulate and the recent development of atomic theory." Como lecture 1927, modified version in Nature 121: 580–590. Danish and English texts in (Bohr 1972), Vol. 6, p. 109–158.

Bohr, N. (1948): "On the notions of causality and complementarity", *Dialectica 2*: 312–318. English text in (Bohr 1972), Vol. 7, p. 330–337.

Bohr, N. (1949): "Discussion with Einstein on epistemological problems in atomic physics", in: Schilpp, P. (Ed.): *Albert Einstein: Philosopher-Scientist* (Library of Living Philosophers, Evanston, Ill.) pp. 200–241. Reprinted in: Bohr, N. (1958): *Atomic Physics and Human Knowledge* (J. Wiley & Sons, New York) p. 32–66, as well as in (Bohr 1972), Vol. 7, 341–394.

Bohr, N. (1972): *Collected Works*, ed. by Rosenfeld, L. and Rüdinger, E. (North Holland, Amsterdam).

Buroker, J.V. (1981): *Space and Incongruence. The Origin of Kant's Idealism* (Kluwer, Dordrecht).

Buroker, J.V. (1991): *The Role of Incongruent Counterparts in Kant's Transcendental Philosophy* in: (van Cleve and Frederick 1991), p. 315–319.

Carnap, R. (1966): *Philosophical Foundations of Physics* (Basic Books Inc., New York). Quoted from the paperback edition (1974): An introduction to the philosophy of science.

Cartwright, N. (1983): *How the Laws of Physics Lie* (Clarendon, Oxford).

Chevalley, C. (1994): "Niels Bohr's Words and the Atlantis of Kantianism" in Faye, J. and Folse H.J. (Eds.): *Niels Bohr and Contemporary Philosophy. Boston Studies in the Philosophy of Science 153* (Kluwer, Dordrecht), p. 33–55.

Chevalley, C. (1995): "On Objectivity as Intersubjective Agreement" in Krüger, L. und Falkenburg, B. (Eds.): *Physik, Philosophie und die Einheit der Wissenschaften* (Spektrum Akademischer Verlag, Heidelberg) p. 332-346.

Cohen, H. (1902): *Logik der reinen Erkenntnis* (Bruno Cassirer, Berlin).

Duhem, P. (1906): *La théorie physique. Son objet – sa structure*. Deuxème édition revue et augmentée (1914), Reproduction fac-similé (Vrin, Paris). (Johann Ambrosius Barth, Leipzig) (New edition 1978: Felix Meiner Verlag, 2nd edition, Hamburg).

Eddington, A.S. (1949): *The Philosophy of Physical Science* (University Press, Cambridge).

Einstein, A. (1905): *Zur Elektrodynamik bewegter Körper. Ann. Phys. 17*: 891–925; reprinted in Lorentz, H.A./Einstein, A. and Minkowski, H. (1982): *Das Relativitätsprinzip*, 8th ed. (Wissenschaftliche Buchgesellschaft, Darmstadt), p. 26–50.

Einstein, A. (1949): "Autobiographical notes" in Schilpp, P.A. (Ed.): *Albert Einstein: Philosopher–Scientist* (Library of Living Philosophers, Evanston, Ill.).

Falkenburg, B. (1993): "The Concept of Spatial Structure in Microphysics" *Philos. Nat. 30*: 208–228).

Falkenburg, B. (1995): *Teilchenmetaphysik. Zur Realitätsauffassung in Wissenschaftsphilosophie und Mikrophysik*, 2nd ed. (Spektrum Akademischer Verlag, Heidelberg).

Falkenburg, B. (1997): "Incommensurability and Measurement" *Theoria 30*: 467–491.

Falkenburg, B. (1998a): "Bohr's Principles of Unifying Quantum Disunities" *Philos nat 35*: 25–120.

Falkenburg, B. (1998b): "Korrespondenz, Vereinheitlichung und die Grenzen physikalischer Erkenntnis" Logos N.F. 5: 215–234.

Falkenburg, B. (2000a): *Kants Kosmologie. Die wissenschaftliche Revolution der Naturphilosophie im 18. Jahrhundert* (Klostermann, Frankfurt am Main).

Falkenburg, B. (2000b): "Bohrs Korrespondenzprinzip und die Grenzen physikalischer Erfahrung" in Enskat, R. (Ed.): *Erfahrung und Urteilskraft* (Königshausen & Neumann, Würzburg, p. 135–147).

Faye, J. (1991): *Niels Bohr: His Heritage and Legacy* (Kluwer, Dordrecht).

Frege, G. (1884): *Die Grundlagen der Arithmetik* (Wilhelm Koebner, Breslau).

Frege, G. (1892a): "Über Sinn und Bedeutung" Z. f. Philos. u. philos. Kritik N.F. 100: 25–50. Repr. in: Frege 1962, p. 40–65.

Frege, G. (1892a): "Über Begriff und Gegenstand" Vjschr. f. wiss. Philos. 16: 192–205. Repr. in: Frege 1962, p. 66–80.

Frege, G. (1962): *Funktion, Begriff, Bedeutung* ed. by G. Patzig (Vandenhoeck & Ruprecht, Göttingen).

Friedman, M. (1983): *Foundations of Space–Time Theories. Relativistic Physics and Philosophy of Science* (Princeton University Press, Princeton, N.J.).

Friedman, M. (1992): *Kant and the Exact Sciences* (Harvard University Press, Cambridge, Mass.).

Giulini, D. et al. (1996): *Decoherence and the Appearance of a Classical World in Quantum Theory* (Springer, Berlin, Heidelberg, New York).

Hacking, I. (1983): *Representing and Intervening. Introductory topics in the philosophy of natural science* (Cambridge University Press, Cambridge).

Hallett, M. (1995): "Logic and Mathematical Existence" in Krüger, L. und Falkenburg, B. (Eds.): *Physik, Philosophie und die Einheit der Wissenschaften* (Spektrum Akad. Verlag, Heidelberg), p. 33–82.

Heisenberg, W. (1942): "Ordnung der Wirklichkeit," unpubl. Reprinted in Blum, W./Dürr, H.-P. and Rechenberg, H. (Eds.) (1985): *Gesammelte Werke, Abteilung C, Allgemeinverständliche Schriften, volume I, Physik und Erkenntnis 1927–1955* (Piper, Munich), p. 217–306.

Heisenberg, W. (1960): "Sprache und Wirklichkeit in der modernen Physik" in: *Gestalt und Gedanke, Vol. 6. Jahrbuch der Bayerischen Akademie der schönen Künste* (R. Oldenbourg, Munich), p. 32–62. Reprinted in Blum, W./Dürr, H.-P. and Rechenberg, H. (Eds.) (1985): *Gesammelte Werke, Abteilung C, Allgemeinverständliche Schriften, volume III, Physik und Erkenntnis 1969–1976* (Piper, Munich), p. 271–301.

Hentschel, K. (1990): *Interpretationen und Fehlinterpretationen der speziellen und der allgemeinen Relativitätstheorie durch Zeitgenossen Albert Einsteins* (Birkhäuser, Basel)

Ishiguro, H. (1990): *Leibniz's Philosophy of Logic and Language* 2nd ed. (Cambridge University Press, Cambridge).

Kant, I. (1755): *Principiorum priorum cognitionis metaphysicae nova dilucidatio* Repr. in (Kant 1910) Vol. I, p. 385-416.

Kant, I. (1768): *Von dem ersten Grunde des Unterschiedes der Gegenden im Raume* Repr. in (Kant 1910) Vol. II, p. 375–383.

Kant, I. (1770): *De mundi sensibili atque intelligibilis forma et principiis* Repr. in (Kant 1910) Vol. II, p. 385–419.

Kant, I. (1781/1787): *Kritik der reinen Vernunft* ed. A/B (Riga: Johann Friedrich Hartknoch).

Kant, I. (1910): *Kant's gesammelte Schriften*, ed. by the Deutsche Akademie der Wissenschaften. (Walter de Gruyter, Berlin).

Mach, E. (1883): *Die Mechanik in ihrer Entwicklung* (Leipzig), 9th ed., 1988 (Wissenschaftliche Buchgesellschaft, Darmstadt).

Mach, E. (1926): *Erkenntnis und Irrtum*, 5th ed. Reprint, 1991 (Wissenschaftliche Buchgesellschaft, Darmstadt).

Newton, I. (1729): *Mathematical Principles of Natural Philosophy* transl. by A. Motte in 1729, ed. by F. Cajori (Berkeley, University of California Press 1934).

Pauli, W. (1994): *Writings on Physics and Philosophy*. Edited by Charles P. Enz und Karl von Meyenn (Springer, Berlin, Heidelberg.)

Plaass, P. (1965): *Kants Theorie der Naturwissenschaft* (Vandenhoeck & Ruprecht, Göttingen).

Planck, M. (1965): *Vorträge und Erinnerungen*, 9th ed. (Wissenschaftliche Buchgesellschaft, Darmstadt), p. 28–51.

Poincaré, H. (1902): *La science et l'hypotheèse*. 1968 (Flammarion, Paris).

Reichenbach, H. (1920): *Relativitätstheorie und Erkenntnis a priori* (Springer, Berlin). Reprinted in: Reichenbach, M. and Kamlah, A. (Eds.): *Gesammelte Werke* (Vieweg, Braunschweig) 1977, Vol. 3, p. 191-302.

Reichenbach, H. (1951): *The Rise of Scientific Philosophy* (University of California Press, Berkeley and Los Angeles).

Scheibe, E. (1997): "Mißverstandene Naturwissenschaft" in Enskat, R.: *Wissenschaft und Aufklärung, Montagsvorträge der Martin-Luther-Universität Halle-Wittenberg* (Leske + Budrich, Opladen), p. 9–29.

Strawson, P.F. (1952): *Introduction to Logical Theory* (Methuen, London).

Suppes, P. (1980): "Größe" und "Messung" in Speck, J. (Ed.): *Handbuch wissenschaftstheoretischer Grundbegriffe, vol. 2 (G-Q)* (Vandenhoeck & Ruprecht, Göttingen), p. 268–269, 415–423.

Tugendhat, E. and Wolf, U. (1986): *Logisch-Semantische Propädeutik* (Reclam, Stuttgart).

van Cleve, I. and Frederick, R. E. (1991): *The Philosophy of Right and Left* (Kluwer, Dordrecht).

Weyl, H. (1949): *Philosophy of Mathematics and Natural Science* (Princeton University Press, Princeton).

8. Idealizations in Physics*

Andreas Hüttemann

Heinrich Hertz was the first to introduce the concept of a symbol to characterize physical knowledge. He invoked this concept to highlight the fact that physical theories are not mere copies of nature. They contain a considerable *constructive* element. As it will turn out the use of idealizations sustains Hertz's claim.

1   An Example of an Idealization

Let us start with an example of an idealization in physics. Assume that we are interested in the question of with what velocity, v, a stone, s, descends in a medium, m, with known viscosity, η. Let us furthermore assume that the stone has an uneven surface. What one usually does in order to solve this problem is to treat the stone as though it were spherical. The stone's actual shape is replaced by a fictitious shape. This is a *conscious* and *voluntary replacement*. It is conscious because we know that the actual shape is not spherical. It is voluntary because we could go on trying to find out the velocity of the stone without replacing its shape – even if this might be more difficult. The replacement allows for an easy calculation of the velocity. We can now calculate the stone's velocity with the help of Stokes' law:

$$F = 6\pi v\eta \Leftrightarrow v = F/6\pi\eta$$

where F is the gravitational force. What we have is illustrated in Fig. 1. The real shape of the stone is replaced by the idealized shape so as to make Stokes' law applicable to it.

2   The Concept of an Idealization

The simple example above allows us to point to three distinctive features of idealizations. First, idealizations are replacements. For reasons that will become clear later on I will be fairly liberal at this point and will allow replacements not only of mathematical descriptions but also of physical systems and data. Thus I take idealizations to be *replacements of either mathematical*

---

* Thanks to M. Adam, H.-J. Glock, T. Kuklinski, J. Kraai, C. Nimtz, B. Priem, J. Rhee and M. Schulte as well as to the other contributors to this book for helpful discussion and suggestions.

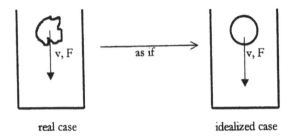

real case   idealized case

Fig. 1.

descriptions, physical systems or data. Second, these replacements are both *conscious* and *voluntary,* i.e., the physicists who idealize have to be conscious of the fact that they idealize. For this reason hypotheses that turn out to be wrong do not count as idealizations. When Galileo proposed his law of free fall, he did not know that his law was wrong. He believed it to be correct. If, however, nowadays a physicist makes use of Galileo's law instead of the correct law based on Newton's or Einstein's theory of gravitation, this has to be classified as an idealization. A hypothesis may *turn out to be wrong;* an idealization is *known to be wrong* (if it concerns theoretical assumptions). It is due to this second aspect of idealizations that the resultant physical theories can be qualified as symbols in the sense of Hertz. Idealizations introduce a constructive element into physical knowledge that is not forced on us by nature. A third characteristic feature of idealizations is that the replacement is not undertaken arbitrarily. The replacement is considered to be *more optimal* in some sense that we have to specify. The focus of the second half of this paper is on the rationale for idealizations and will thus explicate in what sense idealized physical systems, data or descriptions are more optimal than those they replaced.

3   Different Kinds of Idealization

In this section I intend to distinguish various kinds of idealizations that play a role in physical practice. I will start with two kinds of idealizations that physicists will probably refrain from calling idealizations. However, it will turn out eventually that these procedures are closely related to others that are commonly called by this name.

3.1   Production of Physical Systems

Very often the physical systems under investigation are artifacts. They are not part of unmanipulated nature but have rather been produced in factories. Let me quote from an article in which Zeller and Pohl presented the results of measurements of the specific heat and the thermal conductivity of amorphous solids – a case we will investigate in more detail. A table lists the samples that were used in the investigation. For every such entry not only the mass density and the molecular weight is mentioned but also its supplier. The table also tells us how these samples were produced. With respect to one sample it says:

> "The germania sample was melted at 1250°C in vacuum in a Pt crucible, kept at that temperature for 18 h in oxygen at 1 atm, rapidly cooled to 600°C and then slowly to room temperature." (Zeller and Pohl, 1973, p. 2034)

The production of physical systems is an idealization because nature is consciously and voluntarily replaced by artifacts. In what sense artifacts are more optimal than nature will be discussed later.

3.2 Isolation

Physicists typically try to isolate the physical systems they are performing measurements on. Thus the measurement of the specific heat of an amorphous solid takes place in a cryostat. The cryostat is meant to prevent energy exchange between the system and the environment in the low-temperature region. In high-energy physics shielding off unwanted particles plays an important role. E. McMullin has called this procedure "causal idealization":

> "The move from the complexity of Nature to the specially contrived order of the experiment is a form of idealization. The diversity of causes found in Nature is reduced and made manageable. The influence of impediments, *i. e.* causal factors which affect the process under study in ways not at present of interest, is eliminated or lessened sufficiently that it may be ignored." (McMullin, 1985, p. 265)

Isolation is an idealization because a situation in which various causal factors influence the system under investigation is replaced by a situation where no such external factors are present (or less factors). What I presuppose in classifying both production and isolation as idealizations is that physics is a *natural* science, i.e., that it is supposed to deal with natural objects rather than with artifacts. Otherwise these procedures could not be taken to be replacements.

Measurements provide us with data, and it is with respect to data that two further kinds of idealizations need to be mentioned.

3.3 Data Interpolation

Data are typically represented either in tables or graphically as in the following example of the specific heat of various substances (Fig. 2).

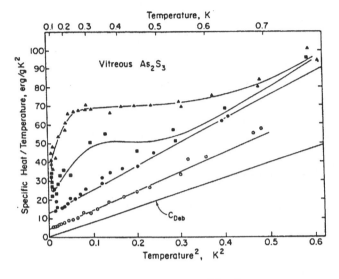

Fig. 2.

The figure is taken from Pohl (1981).

Idealizations come into play because the points that represent the results of measurements are replaced. Duhem has already pointed to the fact that typically a finite number of points is replaced by an infinite number by drawing a curve through the measurement points (Duhem, 1954, Chap. 9). This is what I call *data interpolation*.

3.4   Data Fitting

In general not all of the points that represent measurement results lie exactly on the curve. It is not these points that finally count as the representation of the behaviour of the physical system under investigation. It is rather the curve that is taken to be the phenomenological law. The curve is not meant to capture the exact measurement results. Rather, the interpolation concerns the measurement results and their associated error bars. The error bars are introduced to take into account noise in the data and certain kinds of systematic errors. I call this procedure *data fitting*. Data fitting is an idealization because it involves the replacement of the bare measurement-results by measurement results with error bars.

The phenomenological laws such as those for the thermal conductivity and the specific heat of amorphous solids stand in need of explanation. One would like to know why the behaviour of these systems deviates from ordinary solids. The essential step in the explanation is the construction of a model that can be represented as a Hamiltonian. It is here that two important kinds of idealization are situated.

3.5   Abstraction

Amorphous solids are usually taken to consist of various subsystems: the crystalline structure, electrons and so-called tunneling systems that are responsible for the deviating behaviour of amorphous solids in the low-temperature region. In calculating the contributions of each of these subsystems to the thermal conductivity or to the specific heat, it is assumed that these subsystems are isolated. The crystalline structure's contribution to the specific heat is calculated without taking into account the presence of the other subsystems. That is, the behaviour of the crystalline structure is described in abstraction. The specific heat contribution of the tunneling systems – i.e., its behaviour – is calculated in abstraction as well. The amorphous solid is thus split up conceptually into various subsystems that are treated completely separately from each other. All of these subsystems are described as though they were isolated even though in reality they are part of the amorphous solid and therefore not isolated.

Abstraction is an idealization because the subsystems of a complex physical system are treated as though they were isolated. The description of a subsystem as part of the compound system is replaced by a description of a subsystem that is considered to be isolated. It is assumed that its behaviour can be calculated as though the others were not present (in the absence of interaction).

It should be added that the behaviour of the complex system is usually determined by adding up the contributions of the various subsystems and by taking into account interactions if they occur. It also sometimes happens that not all subsystems of a compound system are taken into account. These are then treated as disturbing factors.

In contradistinction to isolation or causal idealization (Sect. 3.2) abstraction is a purely theoretical procedure, whereas isolation or causal idealization is an operation on physical systems.

3.6   Idealization in the Narrow Sense

It is *idealizations in the narrow sense* which physicists very often have in mind when they employ the concept of idealization: a property of a physical system is replaced by another property that the system is known not to have. Our simple example of an idealization at the outset is an example of an idealization in the narrow sense. Another typical example that plays a role in various areas in physics is the introduction of periodic boundary conditions. In one textbook for solid-state physics we read:

> "A more satisfactory choice is to emphasize the inconsequence of the surface by disposing of it altogether. We can do this by imagining each face of the cube to be joined to the face opposite it, so that an electron coming to the surface is not reflected back in, but leaves the metal, simultaneously reentering at a corresponding point

on the opposite surface. Thus, if our metal were one-dimensional, we would simply replace the line from 0 to L to which the electrons were confined, by a circle of circumference L." (Ashroft and Mermin, 1976, p. 33).

Idealization in the narrow sense is idealization (in the wider sense) because the description of a physical system instantiating some property is replaced by another description of the physical system instantiating another property.

After the construction of a model and the corresponding Hamiltonian, it is in general possible to calculate the physical magnitudes that the system in question ought to have on the assumption that the model is a good model. In the course of this, two kinds of idealization may occur.

3.7   Neglect

Given a certain model, in the course of the calculation certain approximations occur. A typical case of neglect occurs when one has to deal with Taylor expansions. Terms of third or higher order tend to be neglected by physicists.[1] Neglect is an idealization because one mathematical function is replaced by another mathematical function.

3.8   Simplification

A similar kind of idealization occurs when, for Example, in the course of the calculation of a physical magnitude a summation is replaced by an integration. This kind of simplification is an idealization because, as in the case above, one mathematical function is replaced by another mathematical function.

The main point of this list of idealizations is not to claim that it is exhaustive or that all of these various kinds can be distinguished by clear criteria, for example, in the case of simplification and neglect. The aim is rather to give an overview of the scope of procedures that play a role in physical practice. As I have already mentioned, I have been fairly liberal in admitting certain procedures as idealizations that are not usually called so. In what follows I will take those idealizations that are uncontroversially categorized as such as my empirical basis so to speak. Every account of idealizations has to make clear why physicists make use of them. It is *theoretical idealizations* that I take to be uncontroversial idealizations, namely, idealization in the narrow sense, abstraction, neglect and simplification. I will not argue for this classification and take it to be evident. A successful account of idealizations has to give a rationale for all of these procedures. In what follows it will turn

---

[1] Of course in general there are good reasons why certain contributions, e.g. in perturbation expansions, can be neglected. This is, however, not what we are interested in at this point. (See Sects. 5.2 and 5.3 for the relation of empirical adequacy and idealizations.

out that such an account will be able to explain the other procedures I have
included too.

4 Mathematical Simplicity

With regard to the *theoretical* idealizations, i.e., simplification, neglect, idealization in the narrow sense and abstraction, it is not difficult to provide a
provisional rationale for their employment. These procedures lead to simple
theories, and they allow for a simple mathematical treatment of the problems
in question. Thus the authors of a textbook on solid-state physics – Ashcroft
and Mermin – comment on the assumption of a quadratic potential:

> "[it] is not made out of strong conviction in its general validity,
> but on grounds of analytical necessity. It leads to a simple theory –
> the *harmonic approximation* – from which precise quantitative results
> can be extracted."[2]

Similarly in a book on phase transitions the author – Goldenfeld – comments on the use of models:

> "The casual reader of any textbook or research paper on phase
> transitions and statistical mechanics cannot help being struck by the
> frequency of the term 'model'. The phase transition literature is re-
> plete with models: the Ising-model, the Heisenberg-model, the Potts-
> model, the Baxter-model, the F-model and even such unlikely sound-
> ing names as the non-linear sigma model! These 'models' are often
> systems for which it is possible (perhaps only in some limit or special
> dimension) to compute the partition function exactly, or at least to
> reduce it to quadrature (i.e. one or a finite number of integrals rather
> than an infinite number of integrals)." (Goldenfeld, 1992, p. 32).

All of the theoretical idealizations enhance mathematical simplicity. This
is evident in the cases of neglect, simplification and idealization in the narrow
sense. Abstraction, however, helps simple models to be employed as well. It
allows one to split up conceptually compound systems that would have needed
a special treatment. The subsystems can often be described with the help of
simple models.

---

[2] Ashcroft and Mermin (1976), p. 422. The above-mentioned procedure is – strictly
speaking – not an idealization because it is not *known* to be false. It is, however,
not a clear case of an hypothesis either, since it is not invoked because it is
assumed to be true.

5   Idealization and Reality

What has been said so far is rather uncontroversial. Controversy tends to arise as soon as one asks whether idealizations and mathematical simplicity lead to faithful representations of reality. Rom Harré has distinguished two approaches to this question, those of Galileo and of Nancy Cartwright.

> "In Galileo's ontology the methodology of idealizations, expressed in the theorems of geometry, represents actual structures and processes which are the core of reality. (...) On the contrary Cartwright argues that just because the laws of nature express idealizations they cannot be true of the real world. And they express idealization because they are descriptions of idealized models of reality, not messy old world itself." (Harré, 1989, p. 190).

So what we have is two theses, two rationales for the employment of idealizations. According to the Galilean account of idealizations, we invoke these procedures and thereby achieve mathematical simplicity in order to represent reality faithfully. So the ultimate rationale for idealizations is that they lead to a true description of the world. Cartwright argues that idealizations lead us away from truth. In using them we aim at other epistemic values such as explanatory power. Explanatory power and truth do not pull in the same direction in all cases. The rationale for idealizations is that we aim at explanatory power because explanatory power is enhanced by mathematical simplicity (more details below). In those cases where truth and explanatory power move apart, we thus employ idealizations, even though they lead away from a faithful description of reality.

I will argue that neither of these positions can account for all the different kinds of idealization.

Before going into details I need to reject one popular reading of the Galilean account of idealizations.

5.1   Essentialism

One way of explicating the idea that idealizations lead to faithful descriptions is essentialism. The essentialist position has been presented by – among others – by Ellis. He observes:

> "Typically [physical theories] abstract from complex circumstances of nature, and of the imperfections of ordinary physical systems, to consider how ideal systems would behave in ideal circumstances." (Ellis, 1992, p. 265)

He then goes on to give a rationale for idealizations:

> "We do not idealize because nature is too complex to be dealt with without making simplifying assumptions, although this is no doubt

true. We idealize for reasons which have to do with the basic aims of scientific research. Physical science, it will be argued, is fundamentally concerned to discover the essential natures of the kinds of things that can exist, and the kinds of changes that can take place, in a world such as ours. And to achieve its aims, science must focus on the intrinsic properties and structures of the basic kind of things and processes which are to be found existing or occurring in nature." (Ellis, 1992, p. 266)

As examples for "basic kinds of things" Ellis mentions fundametal particles, crystals and stars. It is the essential or intrinsic properties of these things that we are interested in and that we are lead to via idealizations.

"We idealize to remove accidental properties and extraneous forces from centre stage, so we can talk about fundamental intrinsic natures of the kinds of things we are dealing with." (See Ellis 1992, p. 272.)

What essentialism says is that idealizations remove properties of physical systems which are accidental and yield a description of the essential nature of the system in question. A *fortiori* according to essentialism every single form of idealization is a move towards the essential nature of a physical system. However, this is not obvious in some of the cases presented above. Why should we believe that leaving out everything but the quadratic term in a Taylor expansion or replacing a summation by an integration leads towards the discovery of the essential nature of a physical system? At first sight it seems that we use these idealizations for reasons of simplicity rather than for reasons of discovering essential natures. The same is true for idealizations in the narrow sense. Why should we believe that considering a solid as boundless (as in the example in Sect. 3.6) provides us with the description of an essential nature. What Ellis needs to substantiate his claim is an argument according to which the mathematically simpler description leads to the essential nature of a physical system.

Ellis might be tempted to argue as Galileo presumably would have. Galileo believed that the book of nature is written in mathematical language. He therefore took the idealized mathematical description of nature to be the correct description of nature. This is, however, not the problem we have to deal with here. What we need to answer is the question which among various kinds of *different* mathematical descriptions is most likely to be the most faithful description. What Ellis has to presuppose is not only that the book of nature is written in mathematical language but furthermore that the more simple mathematical descriptions are more likely to be true. In the next section I will present an argument to show that this claim is unwarranted. As long as there is no positive argument for such a position it seems reasonable to look for other explanations of why physicists make use of idealizations.

Ellis' position furthermore has to face a problem with abstractions, the case to which his account of idealizations seems to be most suited. This problem has to do with the employment of such concepts as *essential* and *accidental*. The following is how Ellis analyzes why we treat the crystal as though it were isolated in abstraction:

"The crystal structure will almost certainly have flaws and contain impurities of various kinds, which the model will quite properly ignore. The model will ignore these details, because the aim of the exercise is not to save phenomena exactly, but to describe the essential nature of the processes which give rise to the phenomena observed." (Ellis, 1992, p. 276)

In the case of abstraction sometimes one of the subsystems is indeed treated as a disturbing factor and neglected altogether. This, however, is not what happens in general. In the case of a metal the contribution of the crystal and the electrons are completely on a par.

In calculating the contribution of subsystem A to a compound system's behaviour, we abstract from the contribution of subsystem B *and vice versa*. Subsystems A and B are totally on a par. Therefore this kind of idealization cannot be analyzed in terms of essential natures and accidental properties. That would presuppose that one can elevate one and only one of the subsystems of the compound system to the status of an essential nature. Abstraction, however, is completely symmetrical and does not provide the least indication for such an elevation to be legitimate.

5.2   Idealizations and Empirical Adequacy I

The criterion of empirical adequacy will help us to distinguish two classes of idealizations. Physicists make use of these classes for different reasons as will be argued for in what follows.

Let me begin with a quotation from the physicist J.L. Synge. He argues that the use of idealizations is unproblematic as long as the resulting theory is empirically adequate:

"Approximations based on the neglect of small terms are very frequent in mathematical physics, and there is seldom any reason to object to them. One feels that if there is anything wrong, it will show up in some anomaly, and then one can revise the theory." (quoted in Laymon, 1984, p. 115)

Synge indicates that there is a tension between idealizations on the one hand and empirical adequacy on the other (presumably something will "show up in an anomaly" as an empirical *in*adequacy). One will allow for idealizations as long as the discrepancy is not too big. If the discrepancy is felt to be too large tension turns into conflict and the idealizations in question have to be revised. Whether or not a discrepancy can be tolerated is presumably a matter of pragmatic considerations.

The fact that the discrepancies between what is predicted on the basis of idealized models and the phenomena is taken to constitute a tension or in some cases even a conflict indicates that a faithful representation of nature has not been achieved. This is at least what I will presuppose in this paper: empirical inadequacy conflicting with theorizing would point to the fact that the theories or models do not represent reality adequately. It is furthermore the only criterion for unfaithful representation we have. We thus have to conclude that, as long as the tension does not amount to a conflict, those idealizations that lead to mathematical simplicity are invoked *even though* they may lead to a less empirically adequate description, that is to a less faithful representation of nature than a non-idealized description.

This is particularly convincing in the case of simplification and neglect, such as the example of the expansion of the Taylor series, or in the case of idealizations in the narrow sense, such as the case of the boundless solid.

Those idealizations that lead to mathematical simplicity allow simple models to be applied in more cases than in a situation where empirical adequacy were the only thing physicists were interested in.

These idealizations describe physical systems as if they would fall within the range of those simple models that allow for explicit calculation of physical magnitudes. They widen the models' range of application. This is the position Nancy Cartwright advocates in her *How the Laws of Physics Lie*:

> "The aim is to cover a wide variety of different phenomena with a small number of principles [...]. It is no theory that needs a new Hamiltonian for each new physical circumstance. The explanatory power of quantum theory comes from its ability to deploy a small number of well-understood Hamiltonians to cover a wide range of cases. But this explanatory power has its price. If we limit the number of Hamiltonians, that is going to constrain our abilities to represent situations realistically." (Cartwright, 1983, p. 139)

To conclude: We have a rationale for the application of some of the above-mentioned procedures. Idealization in the narrow sense, neglect and simplification allow simple models to be applied to the physical systems. The range of application of these models is enlarged. The fact that a discrepancy between the predictions on the basis of these models and actual measurements is felt to be a tension indicates that Cartwright is right in analyzing these procedures as leading away from the true description of the world.

Widening the range of application of simple models can furthermore be considered to be an explanation for the procedure of data interpolation. Interpolating a certain curve is legitimate as long as no anomaly turns up, i.e., as long as there are not a lot of data far away from the curve. Data interpolation tends to favour simple phenomenological laws, as, for example, laws with integral integers. In general there are established procedures for the explanation of phenomenological laws that vary, say, with the cube of temperature. Simple phenomenological laws allow the range of application of simple models to be widened.

5.3   Idealizations and Empirical Adequacy II

The tension between empirical adequacy and some kinds of idealizations has been taken to indicate that the idealizations in question do not provide a faithful description of nature. There is, however, a kind of theoretical idealization where there is no such tension. This is the case of abstractions. Let me explain this with the help of the following example:

In a paper on amorphous solids the specific heat of amorphous non-metallic solids has been analyzed by the author Hunklinger. Three contributions to the overall specific heat can be distinguished: the crystalline contribution (c_{Debye}), the contribution of the tunneling systems (aT) and a further contribution whose origin was not clear when the paper was written (bT^3).

> "[T]he origin of the 'excess cubic term' bT^3 is less well understood. It cannot be caused by phonons in the ordinary sense because it is known from the acoustic experiments up to 400 Ghz [...] that long wavelength phonons in amorphous solids exhibit hardly any dispersion. On the other hand it cannot be attributed to TS [Tunneling Systems, A.H.] either as we will see [...]." (Hunklinger, 1986, pp. 96–97)

With respect to the specific heat of the compound system (c_V) we thus have:

$$c_V = c_{Debye} + aT + bT^3 \qquad (1)$$

The complex system at hand contains three subsystems: the crystal, the tunneling system and an unknown system. The abstraction I will discuss in what follows is this: on the one hand we consider a subsystem that is itself complex, namely the crystal plus the tunneling system, on the other the unknown system. In calculating the contribution of the crystal plus the tunneling systems, we abstract from the unknown factor. What happens if we test the predictions of our calculations? It turns out that it is empirically inadequate with respect to the physical system under investigation. The term bT^3 quantifies the amount of the empirical inadequacy of the theory of the crystal plus the tunneling systems. Our reaction towards this inadequacy is, however, not to revise the theory we have. Even in the absence of an understanding of the term bT^3 it is legitimate to apply the theory of the crystal plus the tunneling systems to the physical system at hand. Abstraction may lead to empirical inadequacy. The legitimacy of abstraction, i.e., in our case, not taking into account the third term, does not depend on whether bT^3 is small or not. There is simply no tension between abstraction and empirical adequacy as long as there is a reason to attribute the discrepancy to some kind of *further contribution* – whatever that reason might be. Therefore, in the case of abstraction, empirical inadequacy cannot be taken to indicate that representation fails as it did in the case of the above discussed idealizations.

One might object that in both cases, i.e., abstraction on the one hand, and the other theoretical idealizations on the other hand, empirical inadequacy forces us to revise our old theory or to de-idealize. This is true in a certain sense, but the remark does not take into account that the revisions or de-idealizations in the two cases have different implications with respect to the question of whether the old theory was true or a faithful representation of reality. In the case of an empirical inadequacy due to neglect, simplification or idealization in the narrow sense, the idealized (i.e., not yet de-idealized) theory or model has to be revised. It being a false description leads to empirical inadequacy that enforces a revision. In contrast, if we deal with abstractions and it turns out that our model is empirically inadequate, we do not revise the model, we rather add another factor. It turns out that the abstracted model was not the *whole* truth.

This amounts to the following: idealizations such as neglect, simplification and idealization in the narrow sense lead away from truth or a faithful representation of reality, whereas there is no reason to believe that abstraction does. It is thus reasonable to look for a rationale for abstraction in the spirit of the Galilean account.

6   A Rationale for Abstractions

Abstraction separates compound systems into subsystems and tries to understand the behaviour of the compound systems on the basis of the behaviour of these subsystems. C.D. Broad has analyzed various possible kinds of explanations that have recourse to the behaviour of their parts. What is characteristic of explanations in the natural science is what he called "mechanistic explanation" (Broad, 1925, Chap. 2). This notion of mechanistic explanation can best be reconstructed as follows:

A complex system's property can be explained mechanistically if it is – at least in principle – possible to deduce (to explain) the property on the basis of

- (i) the properties of the isolated components,
- (ii) general laws of combination and
- (iii) general laws of interaction.

I do not intend to go into the details of this characterization in this essay.[3] The main point for our investigation is that this kind of explanation requires knowledge about the components' behaviour in isolation. We need to know how the subsystems would behave if they were isolated in order to be able to explain the behaviour of complex or compound systems.

---

[3] For a detailed analysis see Hüttemann and Terzidis (2000).

Abstraction is a procedure that can be understood if one presupposes that physicists make use of it in order to figure out how systems (subsystems) would behave if they were isolated. Knowledge of how systems would behave in isolation is so valuable because there is a very restricted number of laws of interaction and of laws of combination, that allows to calculate all kinds of other systems' behaviour.[4]

It is not only abstraction that can be shown to be a rational procedure on the basis of this assumption. The latter also provides a rationale for why physicists are not only interested in observation but also in experimentation, i.e., in creating phenomena. "To experiment is to create, produce, refine and stabilize phenomena." (Hacking, 1983, p. 230) The best evidence for how physical systems behave if they were isolated is provided through situations in which disturbing factors are absent. The production and isolation of physical systems can be understood as the attempt to realize these conditions. This is why I mentioned these non-theoretical idealizations in the classification in first part of my paper. These procedures are invoked for the same reasons that abstractions are invoked: in order to figure out how systems would behave in the absence of disturbing factors, i.e., if they were isolated.

This rationale also plays a role in explaining data fitting (as opposed to the case of data interpolation). Data fitting is reasonable if one assumes that the deviations are due to unspecifiable disturbing factors. Disregarding these random disturbing influences results in a representation of the isolated system's behaviour.

The knowledge one gains about how the physical systems would behave in isolation, thanks to the procedures of abstraction, production, isolation and data fitting, can then be used to understand the behaviour of more complex systems.

With respect to production, isolation, data fitting and abstraction we have thus proposed a non-essentialist reading of a Galilean account of idealizations. These procedures lead to a true description of the world in so far as they help to discover those components that explain the behaviour of complex systems.

The overall result of our investigation concerning the rationale for idealizations is this: production, isolation, data fitting and abstraction are used in order discover those components that explain the behaviour of complex systems in nature. Idealization in the narrow sense, neglect, simplification and data interpolation are used in order to describe these physical systems such that they fall into the range of application of simple models. We have all reason to believe that the former procedures lead to a faithful representation of what kind of constituents there are in compound systems, whereas the latter procedures provide descriptions of these constituents in terms of simple models.

---

[4] In the case of quantum mechanics it is the simple tensor product rule for non-identical subsystems and its restriction to certain subspaces in the case of identical subsystems.

7   Conclusion: Idealization and Symbol

Idealizations have been defined at the outset as procedures that are invoked willingly and consciously. It is evident that model-building and the calculation of physical magnitudes are constructive activities in the sense that physicists have to decide which procedure to invoke. As has been shown above with respect to idealizations in the sense of Cartwright, there has to be found a balance between empirical adequacy and mathematical simplicity. The construction of models is certainly not determined through the results of measurements alone. Model-building is a theoretical construction process that depends on the active intervention of the physicists. Of course this does not mean that these interventions are arbitrary. As has been shown and as has been pointed out by Hertz, there are criteria that these theoretical constructs have to satisfy. In this sense idealizations in the sense of Cartwright warrant Hertz's claim that physical knowledge is symbolic.[5]

References

Ashcroft, N.W. and Mermin, N.D. (1976): *Solid State Physics* (HRW International Editions, Philadelphia).

Broad, C.D. (1925): *The Mind and its Place in Nature* (Routledge and Kegan Paul, London).

Cartwright, N. (1983): *How the Laws of Physics Lie* (Oxford University Press, Oxford).

Duhem, P. (1954): *Aim and Structure of Physical Theory* (Princeton University Press, Princeton).

Ellis, B. (1992): "Idealization in Science" in: *Idealization IV: Intelligibility in Science*, edited by C. Dilworth (Rodopi, Amsterdam), pp. 265–282.

Goldenfeld, N. (1992): *Lectures on Phase Transitions and the Renormalization Group* (Addison Wesley, Reading Mass.).

Hacking, I. (1983): *Representing and Intervening* (Cambridge University Press, Cambridge).

Harré, R. (1989): "Idealization in Scientific Practice" in: *Poznán Studies in the Philosophy of the Sciences and the Humanities*, 16, (Rodopi, Amsterdam), pp. 183–191.

Hunklinger, S. (1986): "Low Temperature Experiments on Glasses" in J.L. van Hemmen; I. Morgenstern, *Lecture Notes in Physics*, 275, (Springer, Heidelberg) pp. 94–120.

Hüttemann, A and Terzidis, O. (2000): "Emergence in Physics" in: *International Studies in the Philosophy of Science* 14, (Carfax Publishing, Abingdon): pp. 267–281.

Laymon, R.: "The Path from Data to Theory" in: Leplin, J. (ed.) (1984): *Scientific Realism*, (University of California Press, Berkeley), pp. 108–123.

---

[5] For Hertz's view on the notion of symbol in physical knowledge see Chap. 5.

McMullin, E. (1985): "Galilean Idealization" in: *Studies in the History and Philosophy of Science*, 16, pp. 247–273.

Pohl, R. (1981): "Low Temperature Specific Heat of Glasses" in: Philipp, W.A.: *Amorphous Solids, Topics in Currrent Physics 24*, (Springer, Heidelberg), pp. 27–52.

Zeller, R.; Pohl, R. (1973): "Thermal Conductivity and Specific Heat of Noncrystalline Solids" in: *Phys. Rev. B 4*, pp. 2029–2041.

9. Symbolizing States and Events in Quantum Mechanics

Carsten Held

Modern textbooks in statistical mechanics sometimes present classical and quantum-mechanical formalisms in an exactly parallel fashion. At the outset, *pure states* are symbolized by either phase space points or Hilbert space vectors (or, more generally, rays); *observables* by either real-valued functions or self-adjoint operators; and their *values* by either values of the functions or expectation values of the operators. Consequently, also time expansions of states and the values of observables have comparable expressions (see, e.g., Römer and Filk, 1994, p. 47–48). This way of parallelizing the formalisms expresses the general view that they *are* parallel in the sense that there is one set of symbols functioning in the same fashion in both classical and quantum physics, i.e., symbols describing a system's physical *state*, as well as symbols designating *observables* and their *values*.

The procedure does not, of course, intend to conceal the fundamental differences between classical and quantum physics. The classical and quantum physical states are of a fundamentally different kind, insofar as the former determine the values of *all* observables while the latter, notoriously, do not. Quantum physical states fix values only for a proper subset of meaningful observables, and they do so in principle. What they do, in addition, is to yield outcome probabilities for values of all observables. Thus, one might think in the following way: quantum-physical state vectors symbolize physical states like classical phase space points do, in the sense that they are collective symbols of fundamental system properties, but they have the additional feature of yielding probabilities for those values which they do not fix. So, it has become usual, and seems indeed appropriate, to understand the state vector as a "probabilistic system description". The important point, in terms of semantics, is that the basic descriptive function of the symbol, although only for a subset of the observables, is just the same as in the classical phase space point, but there is the *additional* feature of probability, the measurement outcome probabilities derived from the state vector by a straightforward rule of the formalism. The general view is that the fundamental difference between classical and quantum-physical formalism, wherever it is located, lies beyond this semantic parallelism of the basic symbols for system states and properties.

If this picture were altogether correct, we should expect that classical and quantum-physical formalisms, at least in a rudimentary way, start from a common "symbolic platform", before the additional probabilistic feature is brought into play in one of them. But this obviously is false. One formalism operates on real phase spaces, the other on complex Hilbert spaces; one uses points and their motions in phase space, the other vectors or rays and their rotations in Hilbert space. There is, thus, need for clarification of the meaning of notions such as "state" and "event", and there seems need for careful investigation of expressions such as "probabilistic system description".

In the following, I wish to challenge exactly this expression (and, thus, wish to argue that the fundamental difference between classical and quantum physical formalisms does not lie *beyond* but *in* the basic symbols) from a semantic viewpoint, and I wish to clarify the mentioned concepts of "state" and "event". A careful analysis of the semantics of the quantum-mechanical state vector casts severe doubts on the idea of a "probabilistic system description", since, given a straightforward reading of the formalism and some trivial assumptions about probability statements, an incoherence arises which can only be avoided by tacity introducing an extra time index. After presenting the main argument itself, I will try to sketch an alternative conception of the state vector as well as try to clarify, in consequence, the conceptions of "state" and "event" in quantum mechanics.[1]

1   Preliminaries: The Born Rule

I presuppose the normal text book presentation of the quantum-mechanical formalism. Quantum-mechanical states are symbolized as vectors in a Hilbert space dimensioned appropriately for the complexity of the system to be investigated, observables are symbolized as self-adjoint operators acting on the vectors. The time evolution of states is given, for example, by:

$$|\Psi(t_1)\rangle = U(t_1, t_0)|\Psi(t_0)\rangle, \tag{1}$$

where (in the simple case of the system's total energy being time independent)

$$U(t_1, t_0) = \exp[-iH(t_1 - t_0)/\hbar]. \tag{2}$$

The theory also introduces expectation values and presents a statistical algorithm for determining them, for example, from the evolved state vector plus a chosen observable. We are interested only in a very special case, namely the probabilities of certain measurement outcomes, say values a_i of a discrete

---

[1] One word in advance to the purist: Nothing in the following argument hinges on the fact that quantum-mechanical pure states are symbolized by Hilbert space *vectors*. The same purpose could be served for pure states by rays or projection operators, and, for the general case, by density matrices. The semantic points I wish to make in the following only concern pure states – the typical superpositions of eigenstates of an observable; thus I take the liberty to speak of state vectors throughout without qualification. All points about quantum-mechanical states in general – as, for example, the claim that state vectors are mere probability encoders – carry over directly to density matrices in general.

and maximal observable A. The algorithm for these probabilities has come to be called (by philosophers of science) the *Born Rule*. It has the following form:

> BR: If a state $|\Psi(t_0)\rangle$ is prepared, then the probability of finding, upon an A measurement, the system to have value a_k of A is given by

$$\text{prob}(a_k)^{|\Psi(t_1)\rangle}{}_A = |\langle a_k|\Psi(t_1)\rangle|^2 = |\langle a_k|U(t_1,t_0)|\Psi(t_0)\rangle|^2. \tag{3}$$

2 Is the State Vector a Probabilistic System Description? The Main Argument

In the following an argument will be presented to the effect that the state vector cannot, semantically, be understood as a system description; hence, *a fortiori*, it cannot be seen as a "probabilistic system description". The argument is motivated by a remark by Einstein concerning quantum physics. It has been emphasized repeatedly that Einstein's well-known misgivings about the quantum theory did not concern its principled indeterminism (see, e.g., Fine, 1986, p. 26–39, Held, 1998, p. 73–74, 230–39). What he thought, rather, was that quantum physics *structurally* does not conform to a certain ideal of a rational physical description of the empirical world. One way of spelling out this intuition is found in a paper written in 1940. Here Einstein writes, *a propos* the wave function, that it is not at all descriptive, but that, on the contrary, "its only purpose [... is] to make probability predictions". Since the function is the central conceptual tool of the theory, this reduced probabilistic meaning has, in Einstein's view, quite radical consequences for the whole theory:

> "The aim of the theory is to determine the probability of the results of measurement upon a system at a given time. On the other hand, it makes no attempt to give a mathematical representation of what is actually present or goes on in space and time. On this point the quantum theory of today differs fundamentally from all previous theories of physics, mechanistic as well as field theories. Instead of a model description of actual space–time events, it gives the probability for possible measurements as functions of time." (Einstein 1959/1993, p. 109)

Einstein's intuition, thus, is that *because* the state vector's "only purpose is to make probability predictions" the whole theory resigns from "a mathematical representation of what is actually present" or "a model description of actual space–time events". This intuition conflicts directly with the understanding of the vector in most modern interpretations. It is fair to say that nowadays the state vector predominantly is understood as a probabilistic system description (see, e.g., Albert, 1992, p. 30, 35), which effectively means that the tension construed by Einstein between "model description" and a mere encoding of "probability predictions" is seen as a false alternative. In fact, the tacit understanding is that the vector both describes the system *and* encodes measurement outcome probabilities; it has a *double meaning*. This semantic conception of a double meaning for the state vector is refined and made precise in the propensity conception which says that the system properties described by means of the vector are, in general, dispositions or propensities for taking on values of certain observables upon their measurement (see Shimony, 1989, p. 27; Butterfield, 1995, p. 115). If a system is ascribed a state vector, it is thereby said to have (among other properties, perhaps) properties of a certain kind, namely dispositions or propensities for taking on other properties upon measurement. In the following I will argue that this idea cannot be true, if the state vector in fact does, as it must do, encode the Born probabilities. Hence, Einstein's original intuition that there is a radical difference between the state vector interpreted, on the one hand, as a probability encoder function and, on the other hand, as a "model description" of a real system is quite correct.

The argument proceeds in three steps.[2] The *first step* is as follows: The state vector in the standard presentation of the elementary theory (i.e., in the " Schrödinger picture ") carries a time index; this is exactly the presentation chosen above in sect. 9.2. '$|\Psi\rangle$', thus, always is to be spelt out as '$|\Psi(t_1)\rangle$'. Now consider the assumption that the state vector is a system description in the sense of ascribing properties to the system. We can read this in the following way: '$|\Psi\rangle$' is a symbol standing for a collection of system properties; $|\Psi\rangle$ is that collection of system properties. To describe a system by using '$|\Psi\rangle$' is to ascribe to it the collection of properties $|\Psi\rangle$, or, explicitly, is to say, "The system is in state $|\Psi\rangle$." Now, according to the previous reasoning, $|\Psi\rangle$ really is $|\Psi(t_1)\rangle$. What is the semantic function of the time index t_1? The natural choice is that it specifies *at which time* the system is in state $|\Psi\rangle$, so that ascribing $|\Psi(t_1)\rangle$ to a system explicitly means, "The system is in state $|\Psi\rangle$ *at time* t_1." However, *if* $|\Psi(t_1)\rangle$ is understood as a system description, this reading is also the only possible choice by the following argument: The state vector in the Schrödinger picture varies with time. If it is understood as descriptive, then the time at which a certain collection of properties $|\Psi\rangle$ pertains to a system needs to be specified in order to avoid contradiction with other such ascriptions. No reference to time is available apart from the index t_1, hence t_1 must indicate the time at which the system is in state $|\Psi\rangle$.

Consider, in a *second step*, what it means to ascribe to $|\Psi\rangle$ at time t_1 a propensity. A propensity is a quantified disposition to actualize a certain property (see, e.g., Popper, 1982, p. 125–130). The conception of a disposition, in turn, minimally means the possibility of actualizing a non-actual property. I emphasize that I use nothing else but this minimal conception of what a disposition is and that nothing really hinges on any further con-

---

[2] The argument improves a similar one given in Held (1998), p. 249–251.

strual of disposition and/or propensity. Hence, a propensity, at a certain time t_1, of actualizing a non-actual property, entails, by the semantics of modal expressions, that that property-to-be-actualized is, at the very instant t_1, *non-actual*. Trivial examples come from everyday dispositional expressions: A sugar cube which is said at a certain time to be soluble is (by the meaning of "soluble") undissolved at that time; a window-pane which is fragile, i.e. easily breakable, is unbroken. In the same sense, if a quantum system's state $|\Psi\rangle$ at time t_1 entails that the system has a certain propensity to actualize a value a_k of observable A upon measurement, then the system at time t_1 does not have property a_k. Note that the typical cases where propensities play a non-trivial role in the interpretation of the state vector are exactly of this latter kind. For example, in a measurement of an observable A on a system in state $|\Psi\rangle$, where $|\Psi\rangle$ is not an eigenstate of A, the system (since, by the completeness of quantum mechanics, it cannot be said to have any of the values of A, but returns one upon measurement) is said to have a certain propensity to actualize value a_k, which entails that, at the very time when it has that propensity, it does not *have* value a_k as a property. More exactly, it follows from the propensity interpretation that, if $|\Psi(t_1)\rangle$ is not an eigenstate of A, then the system does not have property a_k. Now, by step 1, $|\Psi(t_1)\rangle$ is the state of the system at t_1. Hence it follows, that if $|\Psi(t_1)\rangle$ is not an eigenstate of A, then the system does not have property a_k at t_1.

Now consider, in a *third step*, that '$|\Psi(t_1)\rangle$', symbolizing a pure non-eigenstate of A, also, by BR, encodes measurement outcome probabilities. BR contains only one time index. Thus, intuitively, it seems to say that the square of the time-dependent coefficient $\langle a_k|\Psi(t_1)\rangle$ (the length of the projection of $|\Psi(t_1)\rangle$ onto $|a_k\rangle$) yields the probability for a measurement outcome a_k at time t_1. Explicitly: $\mathrm{prob}(a_k)^{|\Psi(t_1)\rangle}{}_A = \mathrm{prob}(a_k \text{ at } t_1)^{|\Psi(t_1)\rangle}{}_A$. This, however, leads into contradiction. Consider the following trivial principles: (a) If a rule ascribes to an event a non-zero probability, then it must be possible for that event to occur. (b) If, given certain arbitrary premises, it is possible that an event occurs, then, given these premises, it must be admissible to assume that it does occur. Now, consider the case where a system is ascribed a state vector $|\Psi(t_1)\rangle$ which is not an eigenstate of observable A, such that there is, via BR, a non-zero probability for the system having property a_k at t_1. By trivial principle (a), it must be possible that the system has property a_k at t_1. By trivial principle (b), assume that it does have property a_k at t_1. This, however, contradicts what has been concluded, by steps one and two, from the descriptive interpretation of $|\Psi(t_1)\rangle$. Hence, given the trivial principles, the descriptive interpretation must be rejected.

There is not much sense in denying the trivial principles. They express very basic convictions about probability and possibility. The obvious loophole, for adherents of the descriptive interpretation, is to assume that $|\Psi(t_1)\rangle$ does encode probabilities for events happening "a little while" after t_1. This avoids the contradiction in a natural way, but it does so by a tacit modification of BR. If the modification is made explicit, BR now runs:

BRMod: If a state $|\Psi(t_0)\rangle$ is prepared, then the probability of finding, upon an A measurement, the system to have value a_k of A at time t_2 is given by:

$$\text{prob}(a_k \text{ at } t_2)^{|\Psi(t_1)\rangle}{}_A = |\langle a_k|\Psi(t_1)\rangle|^2 = |\langle a_k|U(t_1,t_0)|\Psi(t_0)\rangle|^2. \qquad (4)$$

The descriptive interpretation *requires* such modification, at least tacitly. It is telling, however, that the primitive quantum-mechanical formalism for a simple case like the one considered does not provide any second time index or any obvious handle to bring one into play, theoretically.[3] This fact in itself is a powerful argument against the descriptive interpretation. The bare theory-to-be-interpreted does not talk, in the probabilistic statements produced from the state vector via BR, of more than one time; hence any interpretation doing so tacitly changes the theory, by modifying BR, in order to make sense of it.

But let us consider explicitly which possibilities there are to differentiate t_1 and t_2. One strategy would be to proceed via the peculiar character of quantum-mechanical measurement. The usual way to think about quantum measurement is indeed that, in a non-eigenstate of A, the A measurement interaction starting at t_1 brings about a value of A in the system at t_2 where $t_2 > t_1$. So this picture might well be adduced to justify BRMod. Of course, it is well known that this proposal amounts to a resignation from a certain "classical" conception of measurement, namely that the apparatus registers properties which the measured system has at the beginning of the interaction. The "classical" conception is usually called faithful measurement, and it is generally assumed that quantum measurements cannot be of this kind, at least not in the crucial superposition cases (see, e.g., Redhead, 1987, p. 51–59, 89). Values, are, in fact, not faithfully detected, but rather created in the interaction of system and apparatus.

However, at a point where we are considering a semantic question about the theory, we should abstain from interpretational convictions which arise, given a theory endowed with a full-fledged semantics. Let us, thus, assume nothing else, but the negative fact that quantum measurement is not necessarily faithful measurement. How does this help to justify BRMod? Given that quantum measurement is not faithful, what makes it a measurement at all? In what sense is it informative about micro-systems? The reply here is

---

[3] Note that we could make more time indices explicit, since the *measurement interaction* occupies a certain stretch of time $[t_{begin}, t_{end}]$. This manoeuvre, however, would not do much theoretical work. We will want to say, in the Schrödinger picture, that $t_1 = t_{begin}$ and that t_2 lies somewhere within $[t_{begin}, t_{end}]$. Thus, there possibly is a difference between t_1 and t_2, and nothing else is required and wanted in the present context.

that quantum measurement is to be construed as a *reduced type of measurement*. What quantum measurements merely do is measure *probabilities* (see, e.g., Peres, 1995, p. 24–25). But what does this mean exactly? Very many (ideally infinitely many) measurements of exact replicas of a system in state $|\Psi(t_1)\rangle$ inform us, via the relative frequency of outcomes at times t_1, about the distribution of probabilities for showing values at times t_1, and, via the distribution, to the state vector of an arbitrary element of the ensemble, at time t_1 (see Peres, 1995, p. 25).

Obviously, even in this reduced conception of measurement, nothing indicates a differentiation of time indices of an individual outcome and of the pertaining state vector encoding the probabilities for that outcome. Hence, to reject a certain conception of measurement or to presuppose another is of no help in justifying BRMod. It is, of course, still *possible* to think that frequencies at times t_2 provide information about the state vector at earlier times t_1. But note the amount of interpretational problems this creates. What is the time t_1, if it is different from t_2? Is the time difference the same for all cases in the measured ensemble? Otherwise any conclusions from times t_2 to t_1 would be quite arbitrary. Which mechanism governs the system's evolution between t_1 and t_2? It can obviously not be Schrödinger evolution, nor any effect due to the amplification of the result.[4]

What is so implausible about such manoeuvres is the point where they must be brought into play. They mix ideas from attempts at solving the measurement problem into considerations about a consistent semantics of a basic symbol of the formalism. We should expect that the formalism plus a minimal interpretation of its symbols are minimum requirements for *formulating* the measurement problem. If we are told now that attempts at solving the problem must govern our understanding of a basic symbol of the language in which the problem is set up, we are obviously moving in a circle. It is this fact which gives the whole idea of a double time index its paradoxical appeal. The theory does not endow the state vector with two time indices, and there must be a straightforward and self-consistent interpretation of it, which uses only what the theory explicitly prescribes.

The most powerful argument, however, comes from the fact that the whole idea of a state vector at t_1 encoding probabilities for measurements a little while after t_1 is tied entirely to the Schrödinger picture. In the equivalent Heisenberg picture there is no room for it, which shows that it is an illegitimate addition. In the Heisenberg picture the states are time independent and the operators time dependent. Consequently, time evolution here concerns the observables, e.g. for observable A:

---

[4] Note that these outcomes are *microscopic* properties by definition: value a_k of A is a microscopic property (e.g., a spin component) which cannot be attributed to the apparatus. Thus effects of amplification to the macroscopic, sometimes introduced to solve the measurement problem, play no role here.

$$A(t_1) = U^\dagger(t_1, t_0) A U(t_1, t_0).$$ (5)

And BR here becomes:

BR(H): If a state $|\Psi\rangle$ is prepared at t_0, then the probability of finding, upon an A (t_1) measurement, the system to have value a_k of $A(t_1)$, is given by:

$$\text{prob}(a_k)^{|\Psi\rangle}{}_A(t_1) = |\langle a_k(t_1)|\Psi\rangle|^2 = |\langle U(t_1, t_0) a_k(t_0)|\Psi\rangle|^2.$$ (6)

In the Heisenberg picture the state does not evolve at all. Now, the fact that in the Schrödinger picture the state vector evolves in time was the reason for ascribing it to a system for a certain time between preparation and measurement: to imagine that t_1 is not really the time of the measurement outcome, but the time when the system is in state $|\Psi\rangle$ an instant before it is measured; this picture in turn was the rationale of the double meaning conception of the state vector symbol. In the Heisenberg picture there is not even conceptual room for setting up this picture, since talk of state evolution between preparation and measurement makes no sense here. However, the pictures are entirely equivalent in conceptual content. Hence, the double meaning conception is a superfluous and false addition to the formalism.

The real semantic content of state vector ascription can be seen particularly clearly from the Heisenberg picture, where it is not attached to time expansion. Here, the state vector initially does nothing else but specify the preparation. Then, for an arbitrary time t_1 an evolved observable $A(t_1)$ fixes, together with the prepared state, the probabilities for measurement outcomes *at that time* t_1. The situation is, of course, quite parallel in the Schrödinger picture. Here likewise the state vector does nothing but specify the preparation. For an arbitrary time t_1 the evolved state vector $|\Psi(t_1)\rangle$ fixes, together with a chosen observable, the probabilities for measurement outcomes at that time t_1. Thus, both pictures exemplify that the formalism, read literally, is entirely silent with regard to the system between preparation and measurement. The state vector *a fortiori* does not describe the system between preparation and measurement, but is nothing but a probability encoder function.

3   Two Objections

So far the argument said that the central symbol of quantum mechanics, namely the state vector, cannot, given one minimal semantic function (probability encoding), have another semantic function (system description) at the same time, which it is taken to have in many interpretations of the theory. I have rebutted the most obvious objection, namely that the state vector describes the system "now" and provides probabilities for outcomes "a little while later". However, two further objections deserve discussion.[5]

---

[5] The objections are from Hüttemann and Carrier (objection 1) and Stamatescu (objection 2). I am grateful to them for discussions.

Objection 1 – Dispositions. The argument hinges on an understanding of dispositions as unactualized possibilities, which seems not unexceptional. Surely some dispositions, as, e.g., solubility or fragility, are unactualized possibilities. However, it seems not true that all dispositions are of this kind. If I call a TV set portable, I do so regardless of whether someone actually carries it or not. Hence, dispositions toward certain properties seem independent of the object having or not having these properties. Thus, from the actual pertinency of such a disposition toward a property nothing can be concluded as to the pertinency of that property itself. Might not quantum probability be a disposition of this latter kind?

The objection, I think, rests on a confusion about dispositions. In a very trite sense, statements ascribing possibilities are supported by most empirical property ascriptions. Someone who is (here and now) shot to death, is (here and now) mortal; something which is (here and now) digested, is (here and now) digestible. It is a metaphysical as well as modal triviality that actuality entails possibility. But this is not what we mean, if we, informatively, ascribe dispositions to things or persons. In an informative disposition sentence the speaker implies that, on the basis of certain actual properties, a property is non-actual, but certain or likely to become actual, given that certain additional conditions are satisfied. This is the reason why we intuitively think that someone who says of a precious vase that it is both fragile and shattered commits a semantic mistake. Thus, when someone, pointing at the TV set he actually carries, says that it is portable, we immediately look for a construal of his sentence as informative (Does he mean that the TV set can be carried by anyone, not just him? Or over long distances?). So, the portable TV set puzzle rests on a confusion. We can say, informatively, of the uncarried TV set that it is portable, and we can say of the very same TV set, when carried, that it is portable – but uninformatively. The latter ascription is a purely modal move from the actual to the possible, while the former is an informative disposition ascription. I suggest that we restrict ascription of empirical dispositions (i.e., testable propensities such as probabilities) to the former case.

However, it is not really necessary that we engage in detailed theories of dispositions. The situation envisaged in the argument is the measurement of an observable A where the prepared state is a pure non-eigenstate of A, so that the probabilities for outcomes a_i of A are non-trivial. The completeness of quantum mechanics here seems to prescribe that the system to be measured does not have some value a_k of A which is faithfully measured, but rather that it "takes on" one of the values of A upon measurement (to use von Neumann's memorable phrase; see von Neumann, 1955, p. 206, 211). Thus, the Born probability here is understood as a disposition of the kind mentioned: as a quantified possibility toward actualizing an unactualized value of A. Hence, the conception of dispositions adopted in the argument, and doubted by the objector to be generally applicable, is in play here. Hence, the argument is untouched by any doubts of whether *all* dispositions are actualizations of *unactualized* possibilities.

Objection 2 – Relative-State Interpretations. The contradiction arises from the conflict of a prediction following from BR and the record of the actual measurement outcome. It thus depends on the fact that there is a measurement outcome. Now, of course every reasonable interpretation must admit measurement outcomes, but there is one group – relative-state interpretations – whose adherents claim that there is no fact of the matter as to what outcome as opposed to another possible one is the *actual* outcome; all of them are actual, relative to, for example, a world or a mind (see Everett, 1957; Albert, 1992, Chap.6; Butterfield, 1995; and, for criticisms, Barrett, 1998). Of course, such (many worlds or many minds) interpretations presuppose either an extravagant ontology or a peculiar theory of mind, but they avoid the contradiction, since, according to them, there is no clash between prediction and result. So, the argument is not neutral concerning different interpretations and can be avoided by some.

To see how the objection works in detail, consider the following well-known form of the measurement problem. Suppose, for simplicity, that observable A has only two possible values a_1 and a_2. Suppose that there is an A measurement apparatus which has three possible states: $|\text{"ready"}\rangle$, $|\text{"}a_1\text{"}\rangle$, $|\text{"}a_2\text{"}\rangle$, where the quotes " "$X$" " mean "displays X", such that "$|\text{"}X\text{"}\rangle$" means "a state such that the probability of displaying X equals 1". The apparatus counts as a "good" measurement apparatus if it conforms to the following dynamics:

$$|\text{"ready"}\rangle|a_k\rangle \rightarrow |\text{"}a_k\text{"}\rangle|a_k\rangle \qquad (for\ k = 1, 2). \qquad (7)$$

Consider, now, the typical superposition case where the initial system state is $|\Psi\rangle = \alpha|a_1\rangle + \beta|a_2\rangle$; thus (by the linearity of U):

$$|\text{"ready"}\rangle|\Psi\rangle = |\text{"ready"}\rangle(\alpha|a_1\rangle + \beta|a_2\rangle) \rightarrow \alpha|\text{"}a_1\text{"}\rangle|a_1\rangle + \beta|\text{"}a_2\text{"}\rangle|a_2\rangle. \qquad (8)$$

(Note: The arrow between the left-hand and right-hand sides firstly symbolizes the unitary evolution during the time of the measurement process. In relative-state interpretations it assumes another, peculiar meaning: it symbolizes the branching of the pre-measurement state into a plurality of post-measurement states, attached to a plurality of worlds/minds.) Now, according to relative-state interpretations the right-hand expression does *describe* a post-measurement state, but not a state of one system any more, rather one of many systems in many worlds/minds. Nevertheless, the two components of that state each *describe* an outcome in one world/mind *at the same time* as $|\Phi\rangle$ itself *describes* the total situation. From the relative-state viewpoint this involves no contradiction.

It is not hard to see how the relative-state view evades the argument. The argument presupposes that in the descriptive interpretation of the state

vector the Born probabilities are *propensities*, and in the relative-state interpretation, given that the composite state is actual (simultaneously in many
worlds/minds) and the state components are actual (in one world/mind),
there is no room for any propensity interpretation for the post-measurement
state. Indeed, there is no room any more for any probabilistic interpretation
of the coefficients α and β at all, just because the whole expression and its
components are seen as descriptions. Presumably, α and β either must be
seen as relative frequencies of outcomes avaraged over worlds/minds or have
no meaning at all.

Now, it is a legitimate constraint to require that the symbols have the
same meaning in the left-hand and right-hand expressions. But neither of
the above proposals about what α and β mean seems acceptable. They
cannot, on the left-hand side mean relative frequencies of outcomes distributed over worlds/minds, since before the branching there is no plurality of
worlds/minds, nor are there outcomes. But it is likewise unacceptable to assume that α and β mean nothing at all. This would imply that the Born Rule
be discarded wholesale, together with the possibility of making and testing
quantum-mechanical predictions. The coefficients α and β on the left-hand
side must represent, in one way or other, measurement outcome probabilities,
and, taken together with the descriptive interpretation of the state vector,
they must be propensities. Accordingly, they must have the same meaning on
the right-hand side. Hence, again, the objection does not affect the argument.

4   Quantum States and Quantum Events

Of course, the idea of the state vector being a system description does not
fall out of the heavens. Rather it is seen as the logical consequence of the
completeness of quantum mechanics. Now, as is well known, the completeness
of quantum mechanics can, in a certain sense, be proved mathematically and
tested experimentally. Hence, if it were true that the completeness arguments
show that the state vector for an individual quantum system is a complete
description of that system, we would have produced a veritable dilemma:
on the one hand, the state vector must be a complete description, hence, *a
fortiori*, a description of quantum systems, and on the other hand, it cannot
consistently be a description.

However, such an understanding of the completeness arguments would
be incorrect. What they show is a certain kind of completeness of quantum
mechanics, but not that the state vector is *a complete system description*.
In fact, these arguments are silent with regard to the semantics of the state
vector symbol. What they show is the following: the idea that causal and
dynamical models, hence system descriptions of a certain kind, underlie the
typical quantum situations, together with a small number of extremely plausible additional assumptions, contradicts the quantum-mechanical predictions.
But the impossibility of certain descriptions designed to explain away the

quantum riddles does not entail that quantum mechanics itself or any part of it is descriptive. Naturally, for very many practical purposes the quantum-mechanical state vector can be read, as if it were a system description (see Sect. 9.6 for details). But adducing such cases should not lead us to confuse the foundational semantic questions. At the bottom level, the state vector cannot be a system description, because then it would be contradicting its semantic function of probability encoding. This is what makes precise Einstein's intuition that quantum mechanics does not provide dynamical descriptions (what he calls "model descriptions") of reality. Einstein himself, of course, thought that a fundamental physical theory must provide such dynamical descriptions. Therefore, Einstein reasoned, quantum mechanics could not be a fundamental physical theory. Proponents of a descriptive interpretation share that second intuition with Einstein, namely that a fundamental physical theory must provide dynamical descriptions of physical situations. Since, contrary to Einstein's conclusion, quantum mechanics is a fundamental theory, it is concluded that the state vector is what provides the dynamical description. I suggest, however, that it is Einstein's first intuition, not the second, which should be followed (which is what I have argued for above). Thus, we now need to consider the question of what kind of theory quantum mechanics in fact is, if it does not provide system descriptions. Especially, how are states and events symbolized in quantum mechanics?

Efforts to discard a mistaken semantic conception of the quantum mechanical state vector would be of little interest if they did not lead to a new positive conception. What then, if not system description, is the semantic function of the vector? If the above considerations are correct and are taken seriously, that function is solely probability encoding: the vector is nothing but an encoder for arbitrary probability statements to be specified by choosing an observable. Note what this implies: state vector ascription in itself does not make any sense, strictly speaking. Only a state vector together with a selected observable can lead to concrete probability statements via BR. This connection is appropriately illustrated by the equivalence of the two dynamical pictures, the Heisenberg and Schrödinger pictures. If *only* the specification of a state vector by a preparation, a concrete observable and a concrete time of measurement leads to a full quantum-mechanical proposition, it is inessential whether this statement is obtained by letting the state evolve and attaching the observable to the evolved state or *vice versa*.

The resulting conception is quite reduced. In general, we are given nothing but a continuum of functions encoding probabilities for measurement outcomes. There is, literally speaking, no dynamics of the system in question, and, as argued above, there cannot plausibly be any. There is, of course, something highly counter-intuitive about this picture. At the point of preparation we, apparently, make reference to an individual quantum system, and when ascribing a concrete state vector as the consequence of a specific preparation, what we do is ascribe the vector *to this system*. So we *have* an individual system and some kind of property attached to it which has a time evolution. Doesn't this constitute some kind of dynamics? I claim that, if an individual preparation is at all understood as the ascription of a property to an individual system, then at least a reduced dynamics must be admitted, but, as we will see, even this reduced dynamics leads back into the original difficulty of the double time index. This suggests further that a preparation should not be considered as making reference to an individual system at all.

Consider that a preparation is appropriately characterized by the following statement: "The system is in state $|\Psi\rangle$ at t_0." This will, *via* time evolution plus the choice of an observable A and of a time t_1, lead directly to a collection of probability statements. Thus, we can immediately say: "The system is in a state at t_0, such that for an eventual measurement of A at time t_1, the probabilities of outcomes a_i of A are such-and-such." The system, in the time interval $[t_0, t_1[$, constantly has this property. And this means that the system has *dispositions* for showing certain values of A upon measurement. Even worse, we will run again into the double time index problem. The outcome that is to be given a probability also needs to be time-indexed, and lack of a second index as well as the situation in the (equivalent) Heisenberg picture suggests that that index *is* t_1.

Without the presupposition that an individual system is referred to at the time of preparation, the result of a preparation will be appropriately expressed as: "The prepared state is $|\Psi\rangle$ at t_0."[6] Note that, compared with the previous proposal of a preparation statement, there is an important semantic difference. Whereas in the first form of the statement "$|\Psi\rangle$ at t_0" appeared as a predicate relating to the singular term "the system", so that $|\Psi\rangle$ appears as a property (or collection of properties) at t_0, namely a property of an individual system, in the second form there is no individual to which the state is attached. Accordingly, the first proposal is a predication, while the second states an identity. We could easily transform the second form into the first by replacing "the state" by "the state of the system". In case we wish to understand the state vector as a predicate and the state as a system property, this is in fact what we must do, since we then require an individual system to attach the property to. Here, however, we should resist the temptation and keep the prepared state separate from an individual system to be measured. One may, on the one hand, now accept that states can physically exist without being states of something, thus accept states as self-sustained entities; this seems to be sometimes presupposed by working physicists who talk of the preparation and evolution of certain states rather than systems in certain states. One may, on the other hand, reject this proposal on the grounds that a state, properly speaking, is always a state of something. Even on these grounds, however, it does not follow that the prepared state is necessarily one of an

---

[6] Similar proposals for defining preparations are made by other authors, e.g., Peres (1995) who proposes that a "preparation is an experimental procedure which is completely specified" (p. 12).

individual system. Indeed, here it is appropriate to say, in a Bohrian spirit, that a prepared quantum state is one *of the whole experimental arrangement*.

In accordance with this proposal, the probability statements concluded from the preparation plus the choice of an observable now have a different form: "The state is such that for an eventual measurement of A at time t_1, the probabilities of outcomes a_i of A are such-and-such." Note again the reduced commitment: we can say that the state of the arrangement at time t_0 (the time of preparation) is such that for a measurement at time t_1 the outcome probabilities will be such-and-such. We are not forced into any dynamical picture, however reduced. For sake of consistency it is now necessary, however, to re-interpret the statements expressing measurement results a_i for which probabilities are calculated. We will not express a measurement result as "The system has value a_k of A at time t_1" but rather as "A system with value a_k of A is found at time t_1" in order that any reference to an individual system before measurement is ruled out explicitly.

So far I have argued that accepting the non-descriptive character of the state vector leads us, in a natural way, one step further: The suggestion is that the state vector is not descriptive of an individual system, since it cannot be understood to refer to an individual system at all. Moreover, the argument so far said that, if we do understand the vector as referring to an individual system, there is no plausible way to avoid the descriptive conception. My last step has been to re-formulate those measurement outcome reports for which the theory sets up probabilities. Now, this last step also provides the key to a natural resolution of the double time index problem that brought to grief the descriptive interpretation, but here I have to be necessarily sketchy.

What motivates the idea of a double time index is that identifying the time of measurement (the time when the measurement interaction starts) and the time of the actual outcome (thus the time for which probability statements are made) is equivalent to postulating faithful measurement, and the latter can be disproved by means of quite simple arguments.[7] But these arguments presuppose individual systems to be measured which are identical over different incompatible situations, i.e., situations of different incompatible measurements. Without the assumption of individual systems to be prepared, to be subjected to the prescribed dynamics and to be measured, these arguments collapse.[8] In this reduced case we can assume the time t_1 when the measurement interaction starts and the one for which measurement outcome probabilities are given to be identical. I am unable here to broach the problem of whether this reduced conception can, *in general*, re-install faithful measurement for quantum mechanics, but note that *in the special case*

---

[7] I think here of simplified Kochen–Specker and Greenberger–Horne–Zeilinger arguments such as the ones proposed by Mermin (1990) or, even simpler, a Stern-Gerlach *gedankenexperiment* such as the one proposed by Peres (1995), p. 14–16.

[8] This is pointed out by Peres (1995), again, in his discussion of the Greenberger–Horne–Zeilinger example (p. 153).

it is now consistently possible to say that a measurement at t_1 is a faithful measurement.

Thus, finally, we can concretely formulate what quantum states and quantum events (as opposed to classical states and events) are. A *quantum state* is not the classical state-of-a-system, but the state-of-an-experimental-setup. We remain agnostic about what kinds of microsystems that setup includes, but only give the probabilities for detection of the latter. This new approach to quantum-mechanical states also implies a new conception of events in quantum mechanics. Classically, events are understood as changes of properties in pre-individuated systems. Quantum events – those events for which the theory presents probabilities by means of BR – have been widely understood after the classical model: not, of course, as changes of properties in pre-individuated systems, but as the original acquisition (the "taking on") of properties in pre-individuated systems. Since, in the approach advocated here, the idea of pre-individuated quantum systems is rejected, also quantum events are understood differently. They are not changes of properties in pre-individuated systems, but rather Bohrian phenomena: a *quantum event* is the original appearance of a certain quantum object upon observation.

5   Quantum Mechanics and the Classical World Picture

The resulting picture of quantum mechanics still is so reduced as to appear unacceptable: contrary to the first impression of the formalism (especially the introduction of the state vector and its time expansion), we are not given a dynamics for individual quantum systems, but only continua of probability distributions for finding quantum systems of certain sorts. We encounter individual micro-physical objects, in the models the theory allows as well as in reality, only at discrete instants of time, in the form of quantum events. Moreover, these quantum events themselves, the actual outcomes for which the theory presents probabilities, lie beyond the limits of that theory, at least strictly speaking. This reduced picture of the micro-physical realm is directly opposed to the "new quantum picture" of the world which the formalism seemed to embody originally. Should we not be able to interpret the theory in a more satisfying way, draw a more complete dynamical picture of its physical domain? Is not the reduced picture sketched so far too reduced to do justice to the explanatory power of quantum mechanics?

Two comments are necessary here to defend the approach. First, I have discussed questions of semantics for a minimal and very fundamental piece of quantum mechanics. I have not discussed situations where it is possible to assume stable conditions, follow individual microphysical systems and have dynamical models for their time evolution including continuous ascription of physical properties. Such cases may be found, for example, when systems can be properly isolated, measured repeatedly and can be re-identified. Then, states may be re-interpreted as system properties, and the eigenvalues of these states can be understood as continuous system properties. (Think, for example, of a single atom in a particle trap.) However, strictly speaking, this amounts to a transition from quantum statistics to a classical description, thus to a re-interpretation of the quantum state as a classical state, thus we should not mix such considerations into reflection on the semantics of the former.

Second, if we think that quantum mechanics, simply by virtue of being a physical theory, should embody a family of *dynamical* models of its domain, we follow Einstein in his second intuition where, I suggested, we should follow the first, instead. What bothered Einstein about quantum mechanics was that it presents, in comparison with classical physics, a dramatically reduced world picture. Einstein felt that the theory is dissatisfying just because it is structurally unable to offer dynamical models. It is no accident that in the present course of argument Bohr's elusive views about quantum theory came into play exactly at the point where we wondered how such a reduced picture could be made sufficiently precise. Bohr recommends a modest antirealism and, as far as the dynamics of micro-systems is concerned, even an agnosticism. According to Bohr, we should be aware that, strictly speaking, there is no separate quantum world. There is only one physical world, and quantum mechanics is a necessarily reduced theory of explanation of the quantum-mechanical phenomena.[9] This view has been exploited here in order to make sense of the fundamental symbols of the theory.

References

Albert, D.Z. (1992): *Quantum Mechanics and Experience* (Harvard University Press, Cambridge, MA).

Barrett, J. (1998): "Everett's Relative State Formulation of Quantum Mechanics" in Zalta, E.N. (Ed.),: *The Stanford Encyclopedia of Philosophy.* Available at http://plato.stanford.edu/entries/qm-everett

Butterfield, J. (1995): "Quantum Theory and the Mind: Worlds, Minds, and Quanta". *Proc. Aristotelian Soc. Suppl. Vol. LXIX*: 113–158.

Einstein, A. (1959/1993): *Out of my later years* (Wings Books, New York) (1st edition: Philosophical Library, New York)

Everett, H. (1957): " 'Relative State' Formulation of Quantum Mechanics", *Rev. Mod. Phys. 29*: 454–462.

Fine, A. (1986): *The Shaky Game. Einstein, Realism and the Quantum Theory*, (University of Chicago Press, Chicago).

Held, C. (1998): *Die Bohr-Einstein-Debatte. Quantenmechanik und physikalische Wirklichkeit*, (Schöningh, Paderborn).

Mermin, N.D. (1990): "Simple Unified Form of the Major No-Hidden Variables Theorems", *Phys. Rev. Lett. 65*, p. 3373–76.

---

[9] For more details about Bohr's modest antirealism see, for example, Murdoch (1987), Chap. 10, and Held (1998), Chap. 6.

Murdoch, D. (1987): *Niels Bohr's Philosophy of Physics*, (Cambridge University Press, Cambridge)

Peres, A. (1995): *Quantum Theory. Concepts and Methods*, (Kluwer, Dordrecht).

Popper, K.R. (1982): *Quantum Theory and the Schism in Physics*, (Routledge, London).

Redhead, M. (1987): *Incompleteness, Nonlocality, and Realism. A Prolegomenon to the Philosophy of Quantum Mechanics*, (Clarendon Press, Oxford)

Römer, H.and Filk, T. (1994): *Statistische Mechanik*, (VCH, Weinheim)

Shimony, A. (1989): "Search for a Worldview which can Accommodate our Knowledge of Microphysics" in Cushing, J.T. and McMullin, E. (Eds.): *Philosophical Consequences of Quantum Theory. Reflections's on Bell's Theorem*, (University of Notre Dame Press, Notre Dame, Ind.), p. 25–37.

von Neumann, J. (1955): *Mathematical Foundations of Quantum Mechanics*, (Princeton University Press, Princeton).

10. The Semiotics of "Postmodern" Physics

Hans J. Pirner

1 Introduction

Where does modern physics end? Where does postmodern physics start? The adjective "postmodern" has very special connotations of a new age, different from the previous modern age. Indeed the opinion is expressed that modern physics characterized by the emergence of quantum mechanics and its application to all aspects of microscopic phenomena may be terminating. J. Horgan has given an account of this endzeit in his recent book *The End of Science* (Horgan, 1996). He describes his encounters with great physicists of our times, who give evidence for his hypothesis: "If one believes in science, one must accept the possibility—even the probability that the great era of scientific discovery is over. By science I mean not applied science, but science at its purest and grandest, the primordial human quest to understand the universe and our place in it. Further research may yield no more great revelations or revolutions, but only incremental, diminishing returns."

The physicist A. Sokal has tried to ridicule philosophers who interpret physics in postmodern terms (Sokal, 1996). With a long list of references he gives examples of interpretations of current physical concepts by relativists and social constructivists who emphasize the context in which science is conceptualized. He wrote his article in such a way that the editors of the journal did not realize his hoax and published the text as if it were serious. A transgression of boundaries is a risky enterprise, and any understanding of physical concepts which contains everyday words like "relativity" or "chaos" is bound to lead to interpretations beyond the meaning of these concepts in the physical theories. This is nothing new and occurred before with relativity theory and quantum mechanics.

In fact, M. Beller recently reminded us that the grandfathers of modern quantum mechanics themselves, namely Bohr and Heisenberg, give abundant examples for exporting physics concepts like complementarity to areas like politics or philosophy (Beller, 1998). It seems like a practical joke that they wanted to found an Institute of Complementarity to investigate this concept in all disciplines of human thinking and action. On the contrary, a transdisciplinary approach is a prerequisite in a culture which tries to understand human efforts in the humanities and sciences at the same time. This attempt needs a common vocabulary which suits both enterprises. I propose to explore contemporary physics in semiotic terms. One may debate whether semiotics is a useful tool. Signs and signals are concepts which come from communication theory, a discipline intimately related to telegraphy and electrodynamics, which was invented in the 19th century and is now unthinkable without the chips and computers of the 20th century. So there is some relationship between the philosophical term "sign" and natural science and technology. The symbol concept is more used in the context of language. Symbols are analogues or metaphors standing for some quality of reality that is enhanced in importance or value by the process of symbolization. This chapter will not differentiate strongly between these two terms, and in particular it will use the word "symbolization" also for the semiotic process.

In Sect. 2, I will discuss characteristic new developments in postmodern physics. As examples I have chosen the science of complexity, computer simulations and physical mathematics. I will try to show in which aspects these disciplines go beyond modern 20th century physics. In Sect. 3 the dictionary of communication theory with signs and symbols is introduced. Section 4 interprets the new physics with the help of these concepts and traces the evolution of the language of signs in physics. One could also say it investigates the process of symbolization at its very early stage. Apparently these branches of physics are unfinished, they represent work in progress, which means that their scientific character has not yet been unfolded fully. Therefore, this essay ends with a pragmatist attitude to "wait and see" how these modern fields develop. The philosophical discourse adds awareness, I doubt that it can direct active scientists on how to proceed. A cross-disciplinary dialog which awakens nonscientists to the problematics of scientific progress, however, can improve analytic thinking in the sciences themselves. The possibility of a contract with nature can be established in as far as the perception of nature is concerned. This gives more mutual information to the partners underwriting this contract. M. Serres asks in very romantic words[1] (Serres, 1990; my translation): "How much do we give back to the objects of our science, from where we take our knowledge? Whereas in former times the peasant gave back to the earth via the beauty of his undertaking what he owed to the soil ..." In that sense the semiotics of postmodern physics is not only an epistemological endeavor but also a practical and aesthetic one.

2   Postmodern Fields of Physics

In his book *The Dreams of Reason, The Computer and the Sciences of Complexity*, H.R. Pagels focuses on the science of complexity as the most outstanding new discipline emerging in recent years (Pagels, 1989). M. Gell-Mann, an eminent elementary particle physicist, founded the Santa Fe Institute which is devoted to research in adaptive agent simulation, biological networks, cognition, computational molecular biology, economics, evolving cellular automaton projects, theoretical immunology and neurobiology. All

---

[1] "Que rendons nous, par exemple, aux objets de notre sciende, à qui nous prenons la connaissance? Alorsque le dultivateur, autrefois, rendait en beauté, par son entretien, ce qu'il devait à la terre ..."

these subjects are very complex. The definition of complexity is not easy. "If we try to move towards a mathematical definition, we must realize that the concept of complexity, like entropy, is of probabilistic nature and it can be more precisely defined if we try to define complexity of ensemble of objects of the same category ... ", says Parisi (1988); and he continues in a related article (Parisi, 1994): "The variety of the macroscopic description will be taken as an indication of complexity. An example that is easy to visualize is a heteropolymer, i.e. a polymer composed by a sequence of many different functional units. ... If the polymer may fold in many different ways, we can consider each folding as a different phase and such a system is a complex system." He envisages an ambitious program where in a first step all the possible manifestations of the system can be represented in metric space, i.e. similar configurations classified in clusters, and a tree of such clusters can be constructed, and in a second step the probabilities of the distances in the network of clusters can be calculated. In neural networks physicists have been able to establish a connection between physiological behavior and the dynamics of abstract spins with two states (on and off). The learning rule associates with a small number (p) of patterns a special choice of the coupling matrix between the spin states of different synapses. The system provides associative memory if these p patterns are indeed dynamically stable configurations of the larger system. Also here an ensemble of characteristic pattern states plays a major role (Hopfield, 1982).

P. Anderson, who was one of the strongest opponents of the SSC (Superconducting Super Collider), the biggest accelerator project planned in the US, published his credo in an article with the title "More is Different" (Anderson, 1972), where he claims that all reductionist approaches to nature have a very limited ability to explain the world. All levels are to some degree independent, and each level demands the same creativity and inspiration to be explained as any other. J. de Rosnay says: "Today we are confronted with another infinite: the infinitely complex ... We need a new instrument. As valuable as were the microscope and the telescope in the scientific exploration of the universe. I call this instrument the macroscope. It is a symbolic instrument, constructed from an ensemble of methods and techniques borrowed from very different disciplines." (de Rosnay, 1975). Here a biochemist speaks and one can see the somewhat different perspective. Whereas the physicist adheres to the well-known methods of a mathematical description with or without computers, a scientist of another discipline is more prone to mix methods in order to get a global vision. The physicist prefers the techniques of statistical mechanics of disordered systems, where the system obeying deterministic laws of nature is subjected to a random component. It is hopefully the random component which allows for the variety in the manifestations.

In general, it is more difficult to convey to a young student the importance of a complex system than the importance, of, for example, gravity and cosmology, because the latter disciplines are considered to be fundamental.

They are relevant to our understanding of the universe. If you take a specific macromolecule and its manifestation in a water solution, how does it coil up? Can one attach to different realizations in different solutions a fundamental importance? Can there ever be new fundamental laws in complex phenomena? Note that physics has constructed using statistical mechanics a basic discipline which governs the laws of a large number of particles in large systems. Gell-Mann, the initiator of the Santa Fe Institute, is skeptical about the possibility of discovering similar laws about complex systems. If the term "fundamental" is used to imply expressible in a simple equation or other mathematical calculus, then complex phenomena may not be of that form. Some physicists of complexity have proposed that such systems can only be described by computational codes, where the complexity of the system is related to the length of the code. They claim that complexity is related to the minimal length of the code. The science of machine algorithms goes back to A. Turing (1936) who founded the modern theory of computers. Turing machines are universal machines which combine units for reading and writing codes on different arrays of a storage medium under the control of a processing unit. These Turing machines are extremely simplified theoretical models which help to formulate computations in an organized manner. In this sense also computational approaches to complexity are part of mathematics. It is only in recent years that a coherent attempt has been made to study complex phenomena with experimental and theoretical tools which preserve a holistic view of their components. In the case of methods used to study biological systems, it is especially important that the mechanism of mutual interaction is not obscured by the isolation of the components.

One of the most exciting developments of modern physics are large-scale computations which simulate theories with infinitely many degrees of freedom. After the Second World War, new experimental techniques associated with the development of radar allowed the hydrogen atom to be investigated on a level which is much more accurate than the theoretical description of the atom based on the Schroedinger equation. The electromagnetic field acts not only as a binding potential for the opposite charges, the positive proton and the negative electron, it also modifies the energy levels of the electron, as in a radiation field. Since the fine structure constant (1/137) is a small parameter, the effects of the quantized electromagnetic field are of higher order in the fine structure constant and are calculable term by term.

In contrast, strongly coupled systems are not available for a perturbation theory in a small parameter. Should one therefore give up quantitative predictions? No, if one supplements analytical methods by numerical high-speed computing. Discretizing the world in an artificial lattice of three-dimensional space and one-dimensional imaginary time, one can handle the infinite continuum with a finite number of lattice points. The calculation becomes reasonable once the transition to infinitely many points, i.e. to the continuum, is understood and controllable. Large-scale lattice simulations has become a very important discipline in modern theoretical physics. The building block of the nucleus, the nucleon, is on the verge of being deciphered in this world of bits and strings of code. Not only can quantum phenomena be simulated this way, but thermal fluctuations can also be computed adequately. Modern computers simulate phase transitions where a qualitative change of the symmetry of the system is triggered by varying the temperature. The progress of computational facilities using parallel computers with teraflops operation speeds leads to an improved understanding of many facets of the till-now incomprehensible dynamics of strongly interacting systems. The quantitative change of computing power from the early desktop mechanical calculators to the present-day computers has led to a qualitative change. In a normal numerical calculation each step produces numbers which, after a fixed time and further numerical operations, yield the final result. In numerical simulations, so-called configurations of the system are generated in a probabilistic way and are stored on computer disk as encoded realizations of the system. With the help of these manifestations of the system, more detailed questions can be asked about the mechanism generating the system. Note that the system is produced via a certain prescription. In general, this prescription is simple. The outcome of the simulation, however, is something complicated. Therefore, it may pay to understand it in a different way. In the same way as an experimentalist uses a certain sensor, the computer analyst can add additional code to his simulation to ask pertinent questions about the system which may provide more insight into the dynamics of the strongly interacting system. Let us assume there exists a certain analytical solution of the theory, which we call the "x-ton". This solution may or may not play an important role among the fluctuating quantum realizations of the fields. Now the simulator takes his numerical configurations and checks whether he can identify these pseudoparticles using a filter which eliminates the quantum noise. Some progress has been achieved in this way, but the conclusions are associated with a certain vagueness, since cause and circumstantial evidence cannot be clearly separated.

The development of postmodern physics would have been unthinkable without high-speed computers, a technology which physics has triggered. The other rapid theoretical growth has occurred on the borderline between physics and mathematics. Both of these disciplines have always coexisted and mutually benefited from a vivid exchange of ideas. The common discipline of mathematical physics has developed around this cooperation. Mechanics is associated with the names of, for example, Laplace, Hamilton and Lagrange, and quantum mechanics, i.e. modern physics, is associated with, for example, Hilbert, Weyl and Lie. In postmodern physics the emphasis shifts from physics to mathematics. Whereas historically mathematics has been a tool for solving acute problems in physics, the number of burning problems in parts of physics has been decreasing to a certain degree. Theoreticians have "time off". This is, for example, true of the physics of elementary particles, which has claimed the forefront for a long time. The standard model paired with perturbation theory and numerical lattice techniques has been extremely successful in predicting and explaining the data produced during the last 20 years.

Only the big problem remains, how to unify the hierarchy of different interactions with the weakest interaction, gravity. The enormous progress in the exploration of space and time, using telescopes even beyond Earth, has helped to stimulate interest in cosmology. String theory aims at connecting microscopic elementary particle theory and gravity. It appeared in the late 1960s as an attempt to understand the interaction of protons, and then it hibernated and reappeared in 1984 as superstring theory. This theory lives in 10 dimensions and has a lot of freeway when reduced to our four-dimensional world. Physicists entered the jungle of mathematics to find guiding principles. Two comments have to be made: Once mathematics undergoes axiomatic formulation it brings clarity and transparency. Here, however, we talk about "physical mathematics", conceived during its discovery one may say. The second remark is that the guiding principles for a physical theory are searched for in the Platonic world of mathematics, which is not the case for experimental phenomena. In this spirit, everything which is a beautiful idea will also be realized in nature.

Modern physics conceived point particles as waves, i.e. new quantum mechanical objects when they are studied at microscopic dimensions. Postmodern physics abandons the zero-dimensional point particle, be it wavy or not, in favor of one-dimensional strings, two-dimensional membranes or higher-dimensional p-branes. A trajectory in space–time, called the world line, describes the history of the point particle. Sheets characterize strings propagating and their topology becomes a much more important category than before. The quantum features are built into the theory by the integration over all configurations; in one dimension these would be paths, now they contain the genus, which is the number of handles on the surface of the world sheet. Various divergence problems associated with common field theory now disappear. There is an infinity of string modes corresponding to masses of particles on the order of the Planck scale, which at 10^{-5} g is 10^{17} times larger than the largest masses of the vector bosons. These states contribute as virtual particles to produce subtle cancellation patterns that soften the large momentum behavior of scattering integrals. It is rare in physics that such a giant step in scales can be taken without some other structures appearing. The practitioners in this field explicitly compare their endeavor to the invention of quantum mechanics or to the formulation of the Theory of General Relativity by Einstein in 1916. One must say, however, that the first experimental verification of the predictions of general relativity came in 1919 with the observation of the bending of light rays in the gravitational field during a solar eclipse. The observation of gravitational radiation in a detector is expected in this millenium. Although some historical aspects are similar between the postulate of general relativity and superstring theory, one totally different circumstance is the timescale for when this new theory should come into observational reach. Opinions on this matter are split, but the last 15 years have not seen the goal become closer.

There is another speculative aspect in superstring theories, which is supersymmetry. In models of supersymmetry all the known particles of the standard model possess a partner with a spin reduced by $1/2$. The bosonic photon with spin 1 should be accompanied by the photino with spin $1/2$, which is a fermion. The fermionic quark should have a partner which is a zero-spin particle, the squark. One finds supermultiplets. If local gauge invariance, a feature known from electromagnetic and strong interactions, is combined with supersymmetry, then electric charges and magnetic charges have related strengths, and a relationship can be established between the masses and charges of particles. The mathematical concept of supersymmetry leads to a saturating coupling in the infrared and constrains the quantum corrections to the masses for particles fulfilling the minimal bound.

In this area a spectacular connection to the confinement phenomenon in strong-interaction physics has been established by Seiberg and Witten. The condensation of charged Cooper pairs in superconductivity, which is at work in low-temperature solids, has a mathematical analogue with the condensation of magnetic charge in supersymmetric QCD. Magnetic flux is confined in superconductors, in the dual theory color electric flux, i.e. the quarks are trapped. Here a connection to accelerator laboratory physics appears. The confinement phenomena have experimental starting points, albeit this happens in the supersymmetric theory with more degrees of freedom than in the "real" world. One should not draw the lines of speculation too narrow, we may witness an interesting turning point in physics. It is characteristic that a large number of natural scientists abandon for a significant period the phenomenological world in favor of the world of mathematical ideas. This postmodern development will be analyzed in more detail in Sect. 4.

3   The Semiotics

Historically the concept of sign and symbol goes back to Helmholtz and Hertz (Dosch, 1997). There is nothing postmodern about natural scientists going beyond empirical sensations to abstract information inherent in them. Thus, starting from a physiological basis, the concept of sign as a neural completion of the physical sensation to a meaningful entity was born. As an example, sounds are not perceived as a physicist's analysis would conclude with the intensities distributed over the spectrum given by a frequency analyzer, but the software in our brain develops a sensation of harmony or roughness related to the frequency spectrum. Hertz adds to these perceptions (Hertz, 1894) "our imaginations of the things which have an essential coincidence with the things to fulfill the above explained requirement." This requirement is to produce a chain of symbols (Abbilder) which is related to the chain of events in nature. Hertz introduced symbols which go beyond copies or maps of the physical world into a mathematical universe. These new "signs" become operands by themselves, they enter into chains of "equations" which result in predictions with correspondences in nature. E. Cassirer has elaborated extensively on the concept of symbol, which he sees as the "center and focus of the whole physical science of epistemology" (Cassirer, 1954, p. 25). In general, symbols are more difficult to understand than signs and to define, because unlike signs they are intricately connected to a person or a number of persons sharing the same nationality, civilization or environment. So there is not one lexicon of symbols but many. Signs are more simple, the messages they convey are more mundane. For example, traffic signs have become quite international and have unique meanings. For a down-to-earth analysis of physics they seem to be more useful.

The theory of signs precedes the theory of symbols if one uses C.S. Peirce's text *Syllabus of Certain Topics of Logic* (Peirce, 1993). "A sign is everything which is related to a second thing, which is called its object, in such a way that the sign can determine a third thing, which is called its interpretant, to be related in the same triangular relation to the object, as the sign is related to the object." Next he postulates that this relation is reversible: "This means that the interpretant is a sign by itself, which determines the sign of the (same) object."

The easiest way to come to a concise and clear definition is to use a well-known example in classical physics to explain the terminology and use it to set up the triangle of relations which is so characteristic of semiotics. Take an object like an apple on a tree which is about to fall. The subject calls the apple in front of him the thing to which the sign refers, therefore the object serves as a referent; there may be more than one referent. Studying the distances the apple covers in certain time steps with a fast camera, one can obtain data about the falling apple. If the experimenter is interested in this aspect of the apple, he considers these data as significant data about falling apples. Next he comes to another tree with a different fruit, namely pears and takes similar pictures of falling pears. He compares the coordinate of the traversed distance x with the time t in a graphical plot. If these two plots have a similar parabolic shape, they do not depend on the type of fruit. Now it is useful to speak of apples and pears as something new, say point particles, which obey a law. After some work, which took quite a long time in mechanics, the experimenter may come up with a simple equation of motion. He calls the coordinate $x(t)$ a sign for the position of the massive object above the ground, which is associated with the mathematical equation of motion $d^2x/dt^2 = g$, i.e. the sign is part of a sign language which in the physical sciences is the language of mathematics. In the reverse way the sign determines its interpretants which are the data to be related to the object in the same way as the sign is related to the object. The interpretants cannot add more to the sign than there is already in the sign, they cannot, for example, include data about the temperature of the objects. In its original sense this separation of the sides of the triangle corresponds to the separation into a theoretical and experimental subdiscipline of physics. But one may also apply this separation to higher or lower levels of abstraction. Pierce has built into his epistemological process an infinite regression when he says: "It is essential for the things, that we can only approach them, they can only be represented. The object which a sign wants to represent is a sign by itself." (Pierce, 1986). He enjoys this infinite process and the reflection process which makes his terminology sometimes obscure.

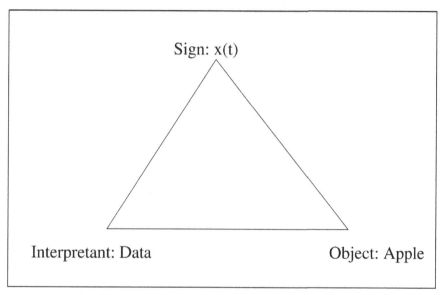

Fig. 1. Triangle representing the different concepts of the semiotic process

With justification, C.S. Pierce can be considered as the founder of semiotics. A philosophical discussion of his work appears in a separate essay by E. Rudolph in this collection. Here, I will only cover some aspects of Pierce's extensive work on signs, introducing some of his terminology and adding my own interpretations as they seem necessary. I will later refer to this discussion in Sect. 4, where a semiotic analysis of "postmodern" physics is attempted. Pierce differentiates between three different types of signs: The simplest type of sign is the "icon". The first view from the ship approaching the harbor in a tropical country shows palm trees, lightly covered people doing their normal activities etc. This is cited by Pierce as an icon of the tropics, and he adds: "All icons from mirages to mathematical equations are similar to themselves, as they do not determine anything, nevertheless they are the sources of all knowledge." In a more prosaic style these icons present a sort of intuitive understanding which precedes a scientific understanding in physics. So, in this sense, geometrical or mathematical constructs belong to iconography as long as the relation of their content to experimental reality is not established. The second type of sign is the "index". It signifies, for example, the place where something can be found in a book. An indicated object is referred to by the index and is put in the context of other objects. So, the relation of the index to the object is more direct than the relation of the icon to the object. The raised "index-finger" suggests a certain direction to the interpretant. In many situations the interpretant is a real person, to whom something is indicated. I do not find it necessary to have persons as intermediaries to interpret natural phenomena in symbolic forms. Mechanically stored data may serve the same purpose, sometimes more objectively. The third type of sign is the symbol which is different from the two other signs, in that it is related to its object solely by the interpretant. Pierce even claims that the symbol determines its interpretant. The symbol conveys a message which depends on convention, usage, or on the natural inclination of the interpretants. In this way symbols may be found in various branches of the humanities, such as literature, history and art. Note, we allow data as interpretants of objects. Data restrict the symbols available for the objects to specific aspects of these objects. The apples have mass but no temperature in the framework of classical mechanics, where we measure time and coordinates.

Pierce has a mystical attachment to the number three. The position of the sign in a threesome, or triad, consisting of "sign–interpretant–object" (Fig. 1) is a characteristic feature in the definition of his semiotics. In his framework, which I support, a dual relationship between the world of objects and the world of mathematical symbols would narrow our understanding of the scientific achievements. We would only see one part of the semiotic triangle, which would present us with the dichotomy between a real and imagined world reflected in the wider context of the philosophy of science under the names "scientific realism" or "social constructivism". In my opinion the historical development of the natural sciences favors a different picture: Masses of empirical data have driven scientific curiosity on a very premathematical basis independently of theories. I call the data interpretants since they give a quantitative picture of the objects and at the same time they interpret the symbols, giving them meaning beyond their positions in a mathematical context. The data connect the level of real objects with the abstract signs making the semiotic triangle complete. If the data can be organized into non-contradictory mathematical symbols, these symbols appear as invariant signs of the objects which are represented by their data. In this respect signs differ from the changing data interpreting different experiments. If there is a law, mathematics will be able to decipher it. Pierce says in *Semiotische Schriften* (Pierce, 1986): "It can be shown to be proven, that no degree of complexity, even if it is infinite, can exceed mathematical imagination."

The threesome or triad of "sign–interpretant–object" can be modified in various aspects. The human interpretant who is outside the triad may enter the triad. Or there are times where the triad develops quasiautomatically. Then the community of scientists become actors who perform a play whose text is prewritten. There are also times when there are interventions, fights and struggles because the semiotic process has become contradictory. T.S. Kuhn has coined the term "scientific revolution" for such changes in the relations of triads. In my opinion, two triads collide with each other. Mechanics and wave theory are in conflict with the description of the same object, the electron. This is not simply a conflict of experiment and theory. It is the whole threesome, the signs and interpretants which differ in relation to the same object.

Other structures emerge when a new triad is built on top of the sign of the original triad. The sign becomes the interpretant of a new triad with a new object and sign. In literature "myth" is such a second-level triad. It treats its low-level abstractions, the words, as interpretants of a narrative. Take the myth about the foundation of Rome. The wolf, a wild unpleasant animal, nourishes Romulus and Remus. The wolf assumes motherlike functions, it transforms itself into a new interpretant signifying the beginning of a civilization out of nature. R. Barthes calls this a shift to a second-order semiological system (Barthes, 1972). Note in this second system an inversion of meaning goes hand in hand with the new position of the "wolf" in the created triad. Barthes continues: "Everything happens as if myth shifted the formal system of the first signification sideways. ... It can be seen that in the myth there are two semiological systems, one of which is staggered in relation to the other: a linguistic system, the language which I call the language object, because it is the language which myth gets hold of in order to build its own system, and myth itself which I call metalanguage in which one speaks about the first." On the second level the meaning of the "wolf" is distorted from wild to motherlike. One can show this shift in a picture (Fig. 2).

Semiology is a developed discipline with many conflicting terminologies, see Eco (1973). At first sight it looks like a schema which then can be applied to almost everything, but does it guarantee deeper understanding? More accurately, the place of the sign in this process starts to rotate and change position from the signified to the signifier. Pierce sometimes uses the index function of the interpretants to point to deeper meaning in the semiotic process. So the active element shifts inside the triad. It is not impossible that also the objects claim more attention than the historical evolution of signification has allowed them.

4   The Semiotics of Postmodern Physics

The semiotic process is very like an expedition without a destination. It is roaming around searching for something. The triad itself is always unfinished.

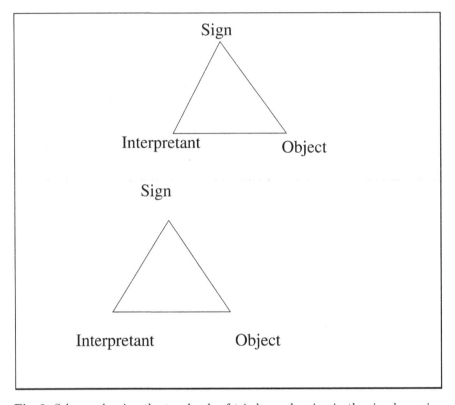

Fig. 2. Schema showing the two levels of triads overlapping in the sign becoming an interpretant

The semiotic process has many features in common with searching the missing corner of the triangle composed of sign, interpretant and object. It is definitely different from scientific research, which is more focused, conscientious and limited. In Sect. 1, I tried to show that many current scientists sense that there may be a significant simplicity beyond apparent complex phenomena. Material sciences in the 20th century started with hard materials, the physics of the solid state. But, more recently, evolution tends towards the soft polymers, soaps, liquid crystals, mixed forms of materials, where order is rarely quantum mechanically determined but by the thermal fluctuations. The theory of random surfaces has made a considerable impact on our understanding of the dynamics of blood cells of tenths of micrometers. Biological objects are envisaged by physicists as referents of significant new data. In my opinion, the science of complexity is mostly engaged in the lower two corners of the triad, gathering possible objects of study and measuring them, i.e. trying to find the key interpretants of these objects. Many experiments have in effect already been done, but the outcome of these series of experiments is so overwhelmingly rich in variety that one speaks of complex phenomena.

Measurement means introducing for these phenomena a new meter stick which allows comparisons between different morphological characters. The categorization of macromolecules may serve as an example. L. Holm and C. Sander have proposed various mappings of molecules to relate protein shapes in a higher-dimensional space (Holm and Sander, 1996). I remind the reader of the work of Parisi cited before. A tree can be established for complex phenomena where, similar to the Linnaean plant classification, the first name is the family name, the second name is the genus, and the last name is the species. When the configurations can be organized into a tree in such a way that the distance between two configurations depends on the position in the tree, the space of the configurations is metric. Complexity can then be defined as a generalization of entropy or neginformation, namely as a double sum of the probability distribution times its logarithm over probabilities and distances in this metric space. It is interesting how the particle physicist G. Mack approaches the same problem using the language of gauge theories (Mack, 1994): "Gauge theory can describe complex adaptive systems, i.e. anything alive in the widest sense, especially autopoietic systems which make themselves in an approximately autonomous fashion."

The sign level for the complex system is each time taken over from another existing field, either condensed matter physics or elementary field theory. An attempt is made to adapt it to a new base of interpretants and referents. In the second case one feels as if one is at the very initial stage of a signification process, in which, even for the practitioner of gauge theory, the analogy is not apparent. It is appropriate to cite E. Cassirer and compare his interpretation of the process of symbolization[2] (Cassirer, 1994, Vol. 1, p. 4; my translation): "Whereas a realistic view of the world ('Weltansicht') rests on a somehow final substantiality of things, as a basis for all cognition, idealism transforms exactly this substantiality into a question of thinking ... Also here (in the individual disciplines of science) the way of thinking does not go from facts to laws and from these laws onward to axioms and fundamental concepts. Axioms and concepts appear at a certain stage as the last and complete expression of the solution, but they must again become a new problem at a later stage. Consequently the object of science cannot be considered any

---

[2] "Wo die realistische Weltansicht sich bei irgendeiner letztgegebenen Beschaffenheit der Dinge, als der Grundlage für alles Erkennen beruhigt, da formt der Idealismus eben diese Beschaffenheit selbst zu einer Frage des Denkens um ... Auch (gemeint ist in den Einzelwissenschaften) hier geht der Weg nicht einzig von den "Tatsachen" zu den Gesetzen und diesen wieder zu den Axiomen und Grundsätzen zurück, sondern eben diese Axiome und Grundsätze, die auf einer bestimmten Stufe der Erkenntnis als der letzte und vollständige Ausdruck der Lösung dastehen, müssen auf einer späteren Stufe wieder zum Problem werden. Demnach erscheint das, was die Wissenschaft als ihr Sein auf ihren Gegenstand bezeichnet, nicht mehr als ein schlechthin einfacher und zerlegbarer Tatbestand, sondern jede neue Art und jede neue Richtung der Betrachtung schließt an ihm ein neues Moment auf."

longer as simple analyzable facts, but each new way or direction of observation opens up a new aspect."

To come back to the subject of biology and complexity, we find that biology has developed a "Weltansicht" for the existence of macromolecules, which is to a large extent focused on the concept of function. Physics has been used as an experimental tool of structural analysis, but can contribute even more insight into the stability and structure of biological forms.

The words of Cassirer cited above sound convincing even to the simple-minded physicist. The practical scientist will be more skeptical reading in the chapter on subjective and objective analysis[3] (Cassirer, 1994, p. 53; my translation): "Thinking experiences its own form through the existence of signs, via the possibility of operating and connecting signs in a specific way following fixed and consequent rules. In this process, thinking reassures itself about its theoretical self. The retreat to the world of signs prepares the decisive breakthrough with which new thought conquers its own world, the world of ideas." Here Cassirer explicitly leaves the method of scientific inquiry, moving into another world of the ideal form "des objektiven Geistes", which one may have problems following.

In another aspect, postmodern physics is involved in a semiotic pattern hitherto unknown in modern physics: large computer simulations of physical theories. Here the transposition of an existing sign in one triad into an interpretant of a new triad occurs. In my opinion, these simulations prepare the shift to a second-order semiological system, fully analogous to the formation of myth in language, as described in Sect. 2. Let me concretize the situation with an example from elementary particle physics: Here large-scale simulations form part of a triad, which includes the proton as an elementary object, a constituent of the atomic nucleus, together with a large class of experiments showing that this proton is composed of quarks and gluons. Currently in physics these quarks and gluons are considered elementary quanta, the dynamics of which are described by a fundamental theory called Quantum Chromodynamics (QCD). This quantum field theory gives us a Lagrangian function which determines the basic equations of motion. The dynamics can also be formulated in computer code and simulated using large number crunchers. The output of these computer calculations is a collection of so-called configurations where the gluon fields have certain values; typically 5000–10000 of these configurations are generated. With these configurations certain properties of the proton can be calculated, for example its mass, or more precisely its mass relative to another elementary particle. This ratio can

---

[3] "An der Form der Zeichen, an der Möglichkeit mit ihnen in bestimmter Weise zu operieren und sie nach festen und durchgängigen Regeln miteinander zu verknüpfen, geht dem Denken jetzt seine eigene Form auf, geht ihm der Charakter seiner theoretischen Selbstgewißheit auf. Der Rückzug in die Welt der Zeichen bildet die Vorbereitung für jenen entscheidenden Durchbruch, kraft dessen der Gedanke sich seine eigene Welt, die Welt der Idee erobert."

then be compared with the experimental ratio and we circle in on a better approximation.

So far so good. There remains the problem of understanding the unfolding of the dynamics, which are entirely formulated by one simple Lagrangian in one line, but the realization after the computer's work goes beyond our intuitive understanding. Here enters the second triad. It consists of approximate pseudoparticles, previously called x-tons, which are analytical solutions of an approximation to the QCD Lagrangian. Perhaps these x-tons can explain the outcome of the simulation? Let's take the gluon field variables as signs of the first triad and work with them. The first triad contains the proton as the object and the measurements about the proton as its interpretants. The signs of the first triad will be shifted to a second new triad where they play the role of an interpretant of this new object, the x-ton; note this is a theoretical object. They will be analyzed in a new program which eliminates certain fluctuations from the original simulation; this may undo quantum effects, and indeed x-tons appear as proposed. One can test whether these x-tons play a significant role by checking whether their presence is correlated with certain properties of the proton, such as the spatial correlation of one of the quarks with the residual quarks. The computer now plays the role of manipulated nature, spitting out a metatheory, i.e. an abstract simplified explanation of the theory of the proton. This formation of "myth" where the original signs become interpretants of a new narrative is quite common in the field of numerical large-scale simulations. Computers are powerful instruments for testing theoretical simplifications, and make the workings of basic physical theories less unattractive for us to consider.

In spite of the simplicity of the underlying Lagrangian, which governs the dynamics in general highly nonlinear strongly coupled field theories, the implications go beyond a simple understanding. A narrative has to be constructed which forms the missing link between the computer and our brain, in the same way that in prelogical times myth mediated between the gods and limited human consciousness.

Large-scale simulations also dominate the more difficult branch of forecasting. "The Limits to Growth", the predictions of the Club of Rome produced in 1972, are an outgrowth of a combination of first-order matrix differential equations with a large number of coefficients; they govern physical growth and decay processes like, for example, in a radioactive decay chain of nuclei (Meadow and Meadows, 1974). Once the coefficients are fitted to previous time histories, the computer extrapolates the solution for the future. In this program there are five main interlocking blocks: population, capital, food, nonrenewable resources and pollution. These influence each other with possible time delays and positive or negative feedback. The method is based on system dynamics, in particular the work of J.W. Forrester (Forrester, 1968). Here the object is a virtual world which exists in the computer. The real world is represented by the input key figures.

The process of symbolization is the modeling of the differential equations, which will be shaped from structural interdependences and then tuned in a repetitive way, i.e. the respective solutions will be examined until some reasonable output data are obtained. In general the output data themselves are not significant, only their interdependences are of value (Meadow and Meadows, 1974): "This process of determining behaviour modes is prediction only in the most limited sense of the word. ... These graphs (i.e. the pictorial results of the model) are not exact predictions of the values of the variables at any particular year in the future. They are indications of the system's behavioral tendencies only." For the empirically minded physicist the triangle is not closed, there is a limited possibility of rejection, gross failures may be visible, and the difference between the virtual world and the real world in simulations is not the same as between an idealized experimental set-up and nature in physics. One talks about computer experiments, because the computer replaces a system in nature or society as the object of our knowledge by a computational schema. We learn more about our possibilities to mimic, to represent the world, but less about how to understand it.

Only in the second step, which I call the semiological shift to a second-order semiological system, when the output data are used as new elements of another triad, do they become interpretants of the real world, with an attached signification which is used to support a new set of beliefs and concepts. This building of the second-level triad is characteristic of the social sciences, where the purely empirical information is mostly insufficient as a trigger for political action. A scenario, i.e. a simulated interpretant of the future, has to be constructed in order to send a strong message. The collapse scenario of the Club of Rome had an incredible impact on the public for the next 20 years.

Postmodern physics examines the limits of scientific inquiry in many other cases. Artificial intelligence and the theory of cognition are other far-out systems which have become playgrounds for physicists. Physical mathematics is more abstract and is aimed at a more profound level. A recent straightforward and simplified introduction to the subject is given by J. Polchinski (Polchinski, 1998). String theory is really a realm of physics, where new mathematical entities are constructed like new "icons". I use the expression "icon" exactly in the sense discussed in Sect. 3, namely as a sign not yet connected to a specific object. Strings or membranes (more precisely noncritical strings) as mathematical objects have their nearest realization in soft-matter theory, like blood cells in biology. Superstring theory does not (yet) have any objects to represent, besides the graviton perhaps. Here physicists are in search of an object. They have the symbolization, they have worked out the iconography for something they do not know. They sense that gravitation may be tightly interconnected to it. But they cannot make the connection.

In order to maintain awareness for something lurking beyond the immediate area of interest, physicists look for bridges to other theoretical signs in other triads. They try to build bridges from the infinitesimally tiny to the infinitesimally small. These would-be bridges extend from string theory to supersymmetric theories and to the Standard Model, which is tested every day at the big laboratories in Chicago and Geneva. Theoretical bridges at the sign level of the icons connect the string icon to the field icons of the Standard Model, which are significant interpretants of data. Here the physicists search for interpretants. The string theory has all kinds of mathematical symbols—what to do with them? In a major archeological effort, relics from the early universe, such as monopoles, strings and domain walls, are searched for. Here the large energy density of the still small universe can compensate for what human-built accelerators cannot yet achieve. This looks again like a search for objects. Note that such searches were successful in the past in the field of elementary particle physics. Purely built on theoretical grounds of renormalizable interactions unifying the weak and electromagnetic phenomena, the postulated W- and Z-particles were indeed found. So such hopes may not be futile. The signs in the mathematics books are leading to the discovery of real things.

The most interesting bridge from these new theories now being constructed aims to include gravitation with the other fundamental interactions. There is now good circumstantial evidence that each of a number of compact X-ray sources in our galaxy contains a black hole of a few solar masses in orbit around a somewhat more massive normal star. On a larger scale there may be black holes of a few thousand solar masses at the centers of globular clusters. When quantum effects are taken into account, black holes are not entirely black, they emit Hawking radiation, which in simple terms is the capture of one part of a particle–antiparticle fluctuation of the vacuum by the black hole, while the other part escapes and appears to be emitted. The black hole is therefore, in general, not a ground state; it will become hotter, radiating its mass away. If the black hole also has a charge associated with it, it will stop radiating when its charge, in suitable units, equals its mass. This type of condition at the extremity corresponds to states in supersymmetric theories, which as BPS (Bogomolny–Prasad–Sommerfield) states also satisfy similar boundary conditions, as discussed before. By a miraculous coincidence it has been possible to calculate the entropy of black holes, i.e. roughly the number of realizations, by counting string states. For the first time this is a link between the up-to-now unattached framework of string signs and the gravitational field. It still presents a puzzle, but shows the far-reaching possibilities in this field.

5 Conclusions

J. Horgan speaks about the ironic mode of doing science in his apocalyptic essay on the end of science (Horgan, 1996): "... to pursue science in a speculative, postempirical mode, that I call ironic science. Ironic science resembles literary criticism in that it offers points of view, opinions, which are best interesting which provoke further comment. But it does not converge on the truth. It cannot achieve empirically verifiable surprises that force scientists to make substantial revisions in their basic descriptions of reality." The protagonists in the fields described above would definitely not consider themselves postmodern ironic physicists. Therefore I put the adjective "postmodern" in quotation marks in the headline of the article.

In this article I have tried to show how contemporary physics has examples of the construction of semiotic processes. The fields of physics discussed are unfinished systems of symbolization, and symbolization is only one of the many aspects of their scientific development. Nevertheless I feel that the study of present-day science injects into the philosophical debate new aspects untouched in a historical analysis. History always separates the successes from the flops. Post facto one may want to know why something succeeded and whether it could not also have failed. Contemporary science is in a disordered state; it presents crossroads, alternatives. The sciences influence our culture indirectly and in a still rather unappreciated way. Here, a dialogue with philosophy may be fruitful. Because of the speed at which modern sciences develop, some of their external interpreters have seen signs of postmodern indeterminism, fragmentation and dissolution. This article does not agree with this categorization. It accepts one property of postmodern thought, however, namely immanence. The scientific process is of this world, and two of the corners of the semiotic triad, the objects and representants, are very much connected to experimentation and data handling, i.e. everyday things. The understanding we presume or gain may finally be connected to other enterprises of culture. The process of symbolization links the natural sciences with language and thought in other fields. It wonderfully illustrates Einstein's remark: "The most incomprehensible thing about nature is that it is comprehensible". To develop a deeper understanding of this question is and will be one of the outstanding tasks in philosophical thinking.

References

Anderson, P. (1972): Science 393, August

Barthes, R. (1972): *Myth Today, Mythologies* (Jonathan Cape Ltd., London)

Beller, M. (1998): The Sokal Hoax, At Whom Are We Laughing? Physics Today, September, pp. 29–34

Cassirer, E. (1994): *Philosophie der Symbolischen Formen*, Vols. I & III (Wiss. Buchges., Darmstadt)

Dosch, H.G. (1997): *The Concept of Sign and Symbol in the Work of Herman Helmholtz and Heinrich Hertz*, Etudes de Lettres, 1-2, pp. 54–61

Forrester, J.W. (1968): *Principles of Systems* (Wright-Allen Press, Cambridge, Mass.)

Hopfield, J.J. (1982): Neural Networks and Physical Systems with Emergent Computational Abilities, Proc. Natl. Acad. Sci. USA 79, p. 2554

Horgan, J. (1996): *The End of Science: Facing the Limits of Knowledge in the Twilight of the Scientific Age* (Addison-Wesley Pub., Reading, Mass.)

Hertz, H. (1894): *Die Prinzipien der Mechanik* (Akademische Verlagsanstalt, Leipzig)

Holm, L., Sander, C. (1996): Mapping the Protein Universe, Science 273, p. 505

Mack, G. (1994): Gauge Theory of Things Alive and Universal Dynamics, hep-lat/9411059

Meadow, D.L., Meadows, D.H. (1974): *The Limits to Growth*, Vol. III (Universe Books, New York)

Pagels, H.R. (1989): *The Dreams of Reason, The Computer and the Sciences of Complexity* (Simon and Schuster, New York)

Parisi, G. (1988): *On Complexity, Measures of Complexity*, ed. by Peliti, L., Vulpiani, A. (Springer, Berlin)

Parisi, G. (1994): *Complexity in Biology: The Point of View of a Physicist*, cond-mat/94120818

Peirce, C.S. (1993): *Syllabus of Certain Topics of Logic* (Phänomen und Logik der Zeichen), transl. by H. Pape (Suhrkamp, Frankfurt)

Pierce, C.S. (1986): *Semiotische Schriften*, Vols. I-III, ed. by Kloesel, C. (Suhrkamp, Frankfurt)

Pirner, H.J. (1990): *Wie und Warum Verändern sich Zeichensysteme? Das Beispiel der Quantenphysik* (unpublished)

Polchinski, J. (1998): *Quantum Gravity at the Planck Length*, Lectures Presented at the 1998 SLAC Summer Institute, Int. J. Mod. Phys. A14: 2633–2658 (1999); hep-th/9812104

de Rosnay, J. (1975): *Le Macroscope vers une Vision Globale* (Editions du Seuil, Paris)

Eco, U. (1973): *Segno* (Istituto Editoriale Internationale, Milano)

Serres, M. (1990): *Le Contrat Naturel* (Editions François Bourin, Paris), p. 68

Sokal, A. (1996): Transgressing the Boundaries: Towards a Transformative Hermeneutics of Quantum Gravity, Social Text 46–47, pp. 217–252